TIME: ITS STRUCTURE AND ROLE IN PHYSICAL THEORIES

SYNTHESE LIBRARY

STUDIES IN EPISTEMOLOGY,
LOGIC, METHODOLOGY, AND PHILOSOPHY OF SCIENCE

VOLUME 179

PETER KROES

Department of Philosophy and Social Sciences,
Technical University of Eindhoven
and
Department of Philosophy, Faculty of Science,
University of Nijmegen, The Netherlands

TIME: ITS STRUCTURE AND ROLE IN PHYSICAL THEORIES

D. REIDEL PUBLISHING COMPANY

A MEMBER OF THE KLUWER ACADEMIC PUBLISHERS GROUP

DORDRECHT / BOSTON / LANCASTER

Library of Congress Cataloging in Publication Data

CIP

Kroes, Peter, 1950–
Time, its structure and role in physical theories.

(Synthese library; v. 179)
Bibliography: p.
Includes indexes.
1. Time. 2. Space and time. I. Title.
QC173.59.S65K76 1984 530.1 84–22255
ISBN-13: 978-94-009-6524-9 e-ISBN-13: 978-94-009-6522-5
DOI: 10.1007/978-94-009-6522-5

Published by D. Reidel Publishing Company,
P.O. Box 17, 3300 AA Dordrecht, Holland.

Sold and distributed in the U.S.A. and Canada
by Kluwer Academic Publishers,
190 Old Derby Street, Hingham, MA 02043, U.S.A.

In all other countries, sold and distributed
by Kluwer Academic Publishers Group,
P.O. Box 322, 3300 AH Dordrecht, Holland.

to *Karin*

Mark

Laura

Thomas

TABLE OF CONTENTS

PREFACE

This book on the structure and role of time in physical theories addresses itself to scientists and philosophers interested in the 'no man's land' between science and philosophy, in particular between physics and philosophy. The point of departure is physical time, i.e. time as used by physicists in their theories; but the analysis is not confined to a purely physical level but caries the problem into the domain of philosophical inquiry. Although the book presupposes some knowledge of physics, I have avoided, wherever possible, the use of advanced mathematics and technical details.

Of all the people who have been of help in writing this book, I would first of all like to mention Paul Scheurer and Guy Debrock who were my primary mentors in science and philosophy. This study is a revision of my dissertation [Kroes 1982a] which I wrote under the stimulating guidance of Scheurer; many of the ideas exposed here have their origin in his work and were developed in frequent discussions with him. Guy Debrock not only stimulated my interest in philosophy but also made valuable suggestions. Without any overstatement, I dare say that without their assistence, this book would never have been written.

Furthermore, I would like to thank D. Dieks, J.J.C. Smart and S. McCall for their comments on the Epilogue and A. Janner and an anonymous referee for their comments on the whole manuscript. My gratitude also goes to all the members of the department of Philosphy of the Faculty of Science of the Catholic University of Nijmegen, and of the department of Philosophy of the Technical University of Eindhoven.

Parts of the material presented here, have already been published elsewhere. I gratefully acknowledge the permission of the editors of <u>Philosophy of Science</u>, of <u>Nature and System</u>, and of <u>Synthese</u> to make use of this published material.

INTRODUCTION

This is a study about 'physical time'. The expression 'physical time' refers to time as used by physicists in their theories; in other words, to time insofar as it plays a role in the description of physical reality. I shall not take into consideration those aspects of time which concern 'psychological time' or 'biological time', nor is it my intention to examine the metaphysical status of time. I shall restrict myself to an analysis of the structure of physical time and of the role time plays in physical theories. This self-imposed 'narrow-mindedness' may be regrettable for several reasons. Most importantly, it leads to a fragmentation in the study of time along the boundaries of established disciplines, while an overall framework for integrating the piecemeal results remains lacking. But it has the obvious advantage of reducing the complexity of the problems since it allows to put aside many questions as not being relevant within the chosen perspective. In the case of a notion as complex as time, such decision is almost unavoidable.

In physics, time is ordinarily introduced by way of a variable t ranging over a certain domain U, the elements of which represent the moments of time, and to which a particular structure is ascribed: the structure of physical time. In most cases, U is taken to be isomorphic to the Euclidean line E; this means that the moments of time have the same order, as well as the same topological and metrical properties as the points on a straight line. This particular choice for the structure of physical time is almost without exception justified by an appeal to our intuitive conception of time: intuitively, time is 'continuous', because a time interval is supposed to be infinitely divisible, has neither beginning nor end, and the moments of time are linearly ordered. The standard idea of physical time, according to which time is isomorphic to a one-dimensional totally ordered continuum, fulfils and combines all these properties of the intuitive picture of time.

This picture of physical time, enthroned in Newton's Principia (1686) as 'absolute time', reigned for about two centuries within physics. Its absolute status was enhanced by Kant's analysis of space and time in his Kritik der Reinen Vernunft (1781) where he argued that space and time are a priori forms of intuition. Admittedly, the ontological status of absolute time was and stayed a topic of controversy which started with the correspondence between Leibniz and Clarke [Alexander 1956], the spokesman of Newton. On the whole, however, the conception of time modeled after a straigth line, was not put into question.

During the 19th century and the beginning of the 20th, several developments took place which strongly stimulated interest in and ultimately revolutionised the physicist's conception of time. In the first place, the discovery early in the 19th century of non-Euclidean geometries destroyed the monopoly of Euclidean geometry and led to the problem of the structure of physical space. This development resulted by the end of that century in Poincarė's conventionalism according to which the structure of physical space is not completely determined by the empirical data alone, but also depends upon certain conventions (e.g. the definition of the physical counterpart of a straight line) [Poincarė 1952, 1958]. Thus, starting from one set of conventions, physical space could be Euclidean, and starting from another set, non-Euclidean. But similar arguments can be and in fact were extended by Poincarė to the structure of physical time. From the conventionalistic point of view the claim that physical time has the structure of a straight line, cannot be justified only on the basis of empirical data; here, too, certain conventions are involved. By choosing a different set of conventions, physical time might very well have a different structure.

Secondly, the interest in the study of physical time was aroused by the study of irreversible processes in thermodynamics. Contrary to the laws of classical mechanics, the second law of phenomenological thermodynamics is, in its standard interpretation, not time reversal invariant; it is valid only relatively to the future orientation of time and not relatively to the past orientation. Therefore it seemed to offer an opportunity to provide a physical foundation to the so-called 'arrow of time' which refers to the 'irreversible process par excellence': the irreversible flow of time from the past toward the future. Henceforth, the problem of the arrow of time has become one of the central problems in the study of physical time.

Last but not least, relativity theory displayed hithertho unknown and unsuspected properties of physical time. In 1905, Einstein showed that our intuitive picture of time, sanctioned in the notion of absolute time, had to be corrected. Moreover, shortly thereafter the introduction in physics of the notion of spacetime, in which space and time merge into a single four-dimensional manifold, opened up completely new perspectives for the study of the problem of physical time.

As a result of these developments, physical space and time became the object of intense study in our century, both by physicists and philosophers. Gradually the 'philosophy of space and time' established itself as a new branch of study within the philosophy of science, more particularly in the philosophy of physics. It deals, among others, with topics like the ontology of space and time, the epistemology of geometry and of time measurement, the global structure of space and time within cosmological models, the causal theories of time (that is the reduction of temporal to causal relations), the physical foundation of the manifold structure of spacetime, irreversibility and the arrow of time.

Three outstanding contributors to this field ought to be mentioned here, not only as an homage to them, but also because they greatly influenced the content of this book:

-Herman Weyl whose <u>Space, time, matter</u>(1921) and <u>Philosophy of mathematics and natural science</u>(1927) have been a constant source of inspiration. From these books especially I learned that physics and philosophy are not disconnected, but that there is a 'no man's land' in between where Weyl was a great pioneer.

-Hans Reichenbach whose <u>The philosophy of space and time</u>(1927) and <u>The direction of time</u>(1956) have become classics; the publication of the first book may well be regarded as the beginning of the philosophy of space and

time as a separate field of research.
-Adolf Grünbaum whose momentous Philosophical problems of space and time(1963) contains a wealth of information.

Originally, it was my intention to write a book about philosophical problems concerning space and time; the philosophical, and especially the epistemological problems pertaining to the structure of physical space in many ways resemble those pertaining to the structure of physical time. After all this is not so surprising in view of the synthesis of both into the notion of space-time. I have restricted myself, however, to physical time. A number of problems which are characteristic for physical time demand a separate treatment, such as the problem of the geometrical nature of time (Ch. II), the problem of time asymmetry (Ch. III, IV) and the problem of the flow of time (Epilogue).

In spite of this restriction to physical time, this study does not cover the entire subject. The ease with which physicists introduce and make use of the concept of physical time in their theories, belies the number and the complexity of the problems to which it gives rise. Indeed, I cannot but agree with Prigogine that "the concept of time is much more complex than we thought" [1980, p. XVI]. That complexity in turn forced me to a further restriction in the range of problems to be treated. Partly for reasons that will soon become apparent, and partly because of personal preference, I have concentrated myself mainly on a systematic exposition of the structure of physical time and the epistemological problems connected with it (Ch. I), on a study of the geometrical role which time plays in physical theories (Ch. II), on the problem of time asymmetry (Ch. III and IV) and on the problem of the physical status of the flow of time. The problem of time asymmetry is more generally known as the problem of the arrow of time, but for reasons to be stated later, I prefer to use the expression 'time asymmetry'. This choice of topics implies leaving out many other interesting aspects of physical time, particularly the problem of its ontological status.

Chapter I contains a systematic presentation of the problem of the structure of physical time. The use of the notion of a differentiable manifold and of the classification of the relations between its elements into different structural levels (the order, topological and metrical level) has been of great help in sorting out the various epistemological problems involved in the determination of the structure of physical time. This mathematical 'thread of Ariadne' has been a reliable guide in finding a way through the labyrinth of literature on this topic.

On the whole, the problem of the structure of physical time is discussed from a conventionalistic point of view. A study of the relation between the structure of physical time and the empirical data upon which this structure is based shows that there exists a gap between both. The empirical data alone do not provide a direct access to the structure of physical time. According to Poincaré's conventionalism, this gap can be bridged, not by collecting ever more data, but only by introducing certain 'free' conventions. A complete determination of the structure of physical time is, in his opinion, possible only on the basis of a combination of these conventions with the empirical data. Therefore any discussion of the 'real' structure of physical time has to remain illusory.

Conventionalism has been accused of leading to relativism; it would imply complete freedom to ascribe any arbitrary structure to physical time provided the corresponding, appropriate conventions are accepted. With regard to such accusations, I would like to point out the following.

-The threat of relativism is rather imaginary. In practice, physicists do not grant each other the liberty to choose whatever structure for physical time. There is general agreement about the basic conventions involved (such as the continuity of physical time); anyone wanting to deviate from these standard conventions would have to produce strong arguments for doing so. Physicists are committed to these conventions because the standard conception of time has proven to be extremely successful in the past. This strong commitment, however, should not be misunderstood; it does not imply that this standard structure represents the 'real' structure of physical time. Indeed, careful study shows irrefutably that conventions play a substantive role. Paradoxically, the freedom of conventionalism does not exclude the pragmatic necessity of a commitment to a particular set of conventions.

-It was certainly not the intention of Poincaré to clear the path for relativism; on the contrary, he was very much opposed to and explicitly rejected that point of view [Poincaré 1958]. His claim that we are free to make a choice between Euclidean and non-Euclidean geometry, has often been misconstrued as the claim that our choice is arbitrary. According to Poincaré this choice is free with regard to the empirical data; in other words, these data do not prescribe which choice to make. But this specific freedom of choice does not imply arbitrariness; we may have good reasons, based upon other criteria, for making a particular choice. Poincaré for instance thought that, with regard to the problem of the structure of physical space, the choice for Euclidean geometry could be justified on the basis of criteria such as simplicity or beauty.
-Instead of looking at conventionalism as leading to relativism, one can also view it more positively as being a kind of antidote against dogmatism in science. I disagree with Lakatos that a 'conservative' conventionalist creates his own prison from which he cannot escape [Lakatos 1978, vol. I, p. 21]. A conventionalist who has made a decision or established a convention does not have to stick to it permanently, but can always reevaluate his choice in the light of new evidence and come back on his previous decision. A hard boiled empiricist, on the other hand, who assumes that the structure of physical time is completely determined by empirical data, is more liable to a dogmatic attitude: once the real structure of physical time has been uncovered on the basis of brute, empirical facts, this structure becomes part of the immutable stock of knowledge which can no longer be questioned. The conventionalist is, or ought at least, always to be aware of the constitutive character of our knowledge of the physical world. Conventionalism conceived in this way, is also more in line with the idea that the scientist creates a physical universe: he does not construct the building of the physical universe with the bricks of empirical data and the cement of logic alone. Conventionalism can free the scientific imagination from the bounds of immutable axioms. Precisely the recognition of the 'conventional' character of one of those seemingly absolute axioms contained for Einstein the clue to relativity theory.

Of course, these remarks do not settle the dispute around conventionalism and relativism; they do not show that the charge of relativism can be refuted. For instance, the criteria of simplicity and beauty invoked by Poincaré, contain their own share of arbitrariness. But these questions are here of secondary importance; my intention is not to give a critical analysis of conventionalism in general, but to show how in the specific case of physical time the conventionalistic point of view manifests itself.

In the second chapter, I shall discuss the geometrical role time plays in physical theories. My interest in this problem was aroused by (i) the simple statement, found in any modern textbook on relativity theory, that since Einstein time has become a coordinate because it transforms under the Lorentz transformations in the same way as the space coordinates, whereas for Newton time functioned essentially as a parameter in the description of the motion of a particle, and (ii) Scheurer's proposal of a theory from which both classical and relativistic mechanics can be derived, depending upon the geometrical role of time. If time is treated as a parameter, this theory leads to classical mechanics; if it is treated as a coordinate, it results in relativistic mechanics [Scheurer 1973].

What is the meaning of this transition from time as a parameter to time as a coordinate? Does it in any way influence the physicist's conception of time or is it only some kind of mathematical trick?

In order to deal with this problem, I have made a detailed analysis and comparison of the 'space and time' description and the 'spacetime' description of the motion of a material (point) particle. In the former, time functions as a parameter for the description of the motion of a particle in space. In the latter it functions both as a coordinate insofar as it is part of the manifold in which motion is represented and as a parameter for parametrising the trajectories of particles in spacetime (in which case it is often called 'proper time'). Historically, the space and time description is connected with classical Newtonian physics and the spacetime description with relativistic Einsteinian physics.

The space and time and the spacetime formalism can be regarded as two different languages for describing physical reality. Of special interest is the question whether or not it is possible to give a translation from one language to the other. A closer analysis of this problem shows that a translation from the space and time to the spacetime formalism can always be given, but that the reverse is more complicated.

Without any doubt, the introduction of the spacetime formalism was fruitful not only for physics itself, but also for the study of physical time (this is why the analysis of the structure of physical time, given in chapter I, is based upon the spacetime point of view). However, the relevance of the distinction between parameter and coordinate time in the spacetime formalism for the study of time has remained almost completely unnoticed. At the end of the second chapter, I shall try to show that this distinction between parameter and coordinate time can throw a new light upon some philosophical problems concerning physical time. Contrary to Reichenbach, for whom the treatment of time was nothing but a convenient mathematical trick relevant to mathematical physics but of no interest for the study of time, I am of the considerate opinion that this distinction can contribute to a better understanding of physical time, in particular of its role in relativity theory.

The second half of this book deals with the problem of time asymmetry, also known as the problem of the arrow of time. Here we touch upon one of the most intriguing but also one of the most obscure aspects of physical time. The complexity of this problem is compounded by a chaotic terminology; beside the expressions 'arrow of time' and 'time asymmetry', one also finds in the literature expressions like 'anisotropy of time', 'direction of time' and 'time orientation', all of which are related to the following cluster of problems: Why does time flow in just one way, from the past toward the future? What is, from a physical point of view, the difference between past, present and future? Is the flow of time an objective phenomenon in the sense that there is a flow of time in physical reality, or is this flow 'mind-dependent'? How can the occurrence of irreversible processes in our universe be reconciled with the time reversal invariant laws of physics? How can we explain the increase of order in our universe? Why can we have traces of the past but not of the future? All these questions are assumed to be related, one way or the other, to the problem of the arrow of time.

At the beginning of chapter III, I shall argue that the unraveling of this cluster requires the preliminary distinction between time asymmetry and flow of time. In a nutshell, time asymmetry refers to the fact that the world looks different relatively to the past and the future orientation of time. Consider the mixing of coffee and cream: the mixing of both components takes place only relatively to the future orientation of time. Looking into the past orientation of time, only the spontaneous unmixing of both occurs. Within the perspective of time asymmetry, all moments of time are on a par; it is not necessary to single out one moment as 'the present' and consequently the distinction between past, present and future is irrelevant for the problem of time asymmetry. With regard to the flow of time, however, the distinction between past, present and future is of primary importance; the flow of time refers to the fact that the near future is turned into the near past by the everchanging now or present. In the standard picture of physical time as a straight line, this flow of time is represented by the motion of a point (the 'present' or the 'now') along the line in the direction of the future. In my opinion, the failure to appreciate the importance of the distinction between time asymmetry and flow of time has caused much confusion in discussions of the problem of the arrow of time.

Chapters III and IV exclusively deal with time asymmetry. The problem of time asymmetry arose in the wake of the 19th century studies of irreversible processes; ever since, time asymmetry and irreversibility have been closely linked together. In fact, it is almost always assumed that the occurrence of irreversible processes is a necessary condition for the occurrence of time asymmetry. I shall start from a more general conception of time asymmetry according to which time asymmetry can also occur with regard to reversible processes. The basic idea underlying the definition of time asymmetry given in chapter III is that time asymmetry exists with regard to processes of a certain kind A in case the probability for the occurrence of a process A differs from the probability for the occurrence of the reversal of A relatively to the same orientation of time. Time asymmetry, thus defined, is a probabilistic phenomenon.

In order to determine the connection between time asymmetry and the time reversal invariance or non invariance of physical laws, the nature of the time reversal operator T* will be discussed. It will be shown that in certain cases the action of T* on physical quantities can be defined coherently in different ways. This again has important consequences for what is meant by the time reversal of a process, for the reversibility or irreversibility of a process and for the time reversal invariance or non invariance of physical laws; depending upon the definition of the action of the time reversal operator T* on physical quantities, a law may be time reversal invariant or not.

The analysis of the notion of time asymmetry and of irreversibility in chapter III, sets the stage for the final chapter. There, the origin of one particular kind of time asymmetry is discussed, namely the time asymmetric behaviour of thermodynamical systems: a closed system not in equilibrium (in a relatively low state of entropy) evolves toward equilibrium (to the state of maximum entropy) but the evolution in the opposite way, from equilibrium to non equilibrium, never takes place spontaneously. This time asymmetric behaviour is captured in the second law of phenomenological thermodynamics which states that the entropy of a closed system increases or stays equal, but never decreases.

In chapter IV, two opposing views on the origin of thermodynamic time asymmetry will be discussed. The standard conception of thermodynamic time asymmetry, referred to as the 'mechanical program' and connected among others with the names of Boltzmann, Ehrenfest, Reichenbach and Grünbaum, holds that thermodynamic time asymmetry is the result of certain (accidental) boundary conditions prevailing in our universe; in other words, thermodynamic time asymmetry is supposed to be of a de facto nature. The second law of phenomenological thermodynamics, which is incompatible with the laws of mechanics, is regarded to be a 'second rate' law which is only valid within a macroscopic description of physical reality. It cannot be strictly valid in its original form, but must be replaced by the so-called statistical H-theorem.

The 'thermodynamical program' with regard to the problem of the origin of thermodynamic time asymmetry, on the other hand, holds that the second law is of the same fundamental nature as the time reversible laws of mechanics. This position has been advocated by, for instance, Planck, Stueckelberg and Prigogine. According to this view, thermodynamic time asymmetry finds its origin in one of the fundamental laws of nature and not in boundary conditions. In view of the incompatibility of the second law with the laws of mechanics, this conception of thermodynamic time asymmetry makes it necessary to reconsider the status of mechanics; a way must be found to combine the irreversible, thermodynamic description of physical reality with the reversible, mechanical one.

The mechanical conception of thermodynamical time asymmetry has become dominant in physics. Only occasionally, the thermodynamical program is defended. Recently, Stueckelberg and Prigogine have given a new push to this program by developing, each in his own way, an 'irreversible physics'; in their theories, the second law of phenomenological thermodynamics emerges as one of the most fundamental laws of nature. Especially Prigogine has strongly attacked the mechanical interpretation of thermodynamical time asymmetry. His arguments against the mechanical program will be examined in detail.

In most discussions of the problem of the origin of thermodynamical time asymmetry, the ideas developed within the thermodynamical program are neglected. There is, however, no reason to reject this program beforehand. That is why I have undertaken an examination and comparison of both programs in chapter IV, where the most significant differences between both programs will be pointed out.

In the epilogue, I turn to a discussion of the flow of time. Whereas time asymmetry is generally regarded to be a 'legitimate' physical concept, the status of the notion of flow of time within physics is more controversial; the question is whether or not physical time flows. In other words, is the phenomenon known as the flow of time part of the structure of physical time? Very often, these questions are answered in the negative; the impression of a flow of time, it is said, is nothing else but an illusion which is not relevant for the description of physical reality and therefore the flow of time is not part of the structure of physical time. This view holds that concepts such as 'the past', 'the present', 'the future' and 'the flow of time' (or 'becoming') are not genuine physical concepts since they only acquire some meaning within the context of the experience of a conscious observer; they are 'minddependent' concepts. They are empty concepts as far as they pretend to say something about physical reality itself. If this were true, a discussion of the flow of time would fall outside the scope of this book.

Yet, it is paradoxical that such a fundamental aspect of our awareness of time should not be incorporated into the structure of physical time. Therefore attempts have been made to give an objective, physical foundation to the flow of time. The aim of objective theories of time flow is to state within a purely physical language (which means that any reference to a conscious observer must be avoided) an intrinsic difference between events belonging to the past, the present and the future. The underlying idea is that there is a flow of physical time and that 'becoming' is an aspect of physical reality.

The epilogue contains a discussion of the minddependent and objective theories of time flow. A closer look at both points of view shows that they are beset by conceptual difficulties. On the one hand, the minddependent theories of time flow lead for instance to the rather awkward conclusion that, from a physical point of view, the only difference between time and space consists in their number of dimensions. The objective theories of time flow, on the other hand, appear to strive for the impossible, namely to single out one unique moment of time - the present - on purely physical grounds, whereas physics in general disregards that which is unique in phenomena and is based upon the reproducible.

Whether or not it is in principle impossible for physics to incorporate the flow of time in its description of physical reality, is still an open question. Up to the present, all attempts to capture this mysterious but essential aspect of time in the language of physics have failed.

CHAPTER I

PHYSICAL TIME AND THE PROBLEM OF ITS STRUCTURE

I.1 Introduction

This chapter deals with the structure of physical time. Within classical Newtonian physics, with its strict separation between space and time, the meaning of this expression 'the structure of physical time' is rather unproblematic: time, as conceived by Newton, is an entity existing by itself, per se, and this entity has certain properties. The whole set of these properties ascribed to time will be referred to here as 'the structure of (physical) time'. Of special interest is the fact that this structure belongs to time itself, regardless of the relation time bears to space or any other physical quantity, or of the way time is being used in physical theories; as is well known, time, within a relativistic context, looses much of its independence, especially with regard to space.

Within a classical context, therefore, the question "What is the structure of time?" is meaningful, and the answer runs somewhat as follows: time has the structure of an Euclidean line along which the 'now' or the 'present' is moving constantly toward the future. Apart from this motion of the now, the structure of time is isomorphic to the structure of a straight line. For Newton it was not possible to think of time in another way than as a straight line. Therefore time had neither a beginning nor an end, and it was infinite toward the past and the future. Moreover, the moments of time were ordered like the points on a line, thereby eliminating the possibility of a cyclic time. This image of time, as a straight line, determined all its physically relevant properties.

Within relativistic Einsteinian physics, however, the notion of time itself, and a fortiori of the structure of time, becomes problematic. Time and space can not be completely separated, as is most clearly demonstrated in the by now standard spacetime approach to relativistic physics (1). Within the latter, spacetime is taken to be the fundamental spatio-temporal entity. Time as such no longer exists, nor does space [Minkowski 1908]. The temporal relationships between events depend upon the spatial ones and vice versa. In the course of this chapter, the precise nature of this relationship between space and time will be elucidated. For the moment, the issue whether we can in a sensible way speak of the structure of time in a relativistic context is more important.

Suppose that we had at our disposal a unique way of splitting up an entire spacetime manifold into two components, space and time; i.e., given a spacetime manifold M, we apply a certain procedure S which yields M as the direct product of a three-dimensional and one-dimensional manifold, representing respectively space and time. Now, if for each and every observer the result of applying S to M is the same, in other words, if for each observer the procedure S yields the same separation of spacetime into space and time, the use of the notion 'time' and of 'structure of time' appears to be justified. Although time is thus constructed from spacetime, it nevertheless makes sense to speak of (the structure of) time. Like for Newton, time has in that case a universal character: the properties of time are the same for all observers, independently of their place and state of motion.

Unfortunately, however, there exists no such procedure S which produces a universal separation of a relativistic spacetime M into space and time, because temporal relations between events are observer dependent [Weyl 1952, p. 130 ff]. For instance, it is impossible to split up an arbitrary spacetime M into space and time so that it incorporates the simultaneity of events according to two observers in relative motion. The same applies to the measurement of time intervals by observers in relative motion. Moreover, within a given spacetime manifold, the global properties of time, e.g. whether it is closed or open, may vary from one observer to the other.

For these reasons, 'time and space talk' within a relativistic context is highly problematic. The notions 'time' and 'the structure of time' have at first sight lost much of their significance. However, this should not be exaggerated; it does not mean that these notions cannot be used anymore in a sensible way within a relativistic context.

Indeed, the use of the notions of time and of space in a global sense requires great caution. But, on the other hand, locally any spacetime manifold looks the same for any observer, independently of his state of motion, viz., like Minkowskian spacetime. Therefore the temporal aspects of this Minkowskian spacetime are of a general nature; they are valid for all observers. In this sense, time has locally the same structure for all observers, in spite of the fact that the temporal relations between events are observer dependent. Thus locally the notion of time remains meaningful. Problems occur when we go from local time to global time. Indeed, the notion 'global time' itself is highly problematic, especially in those cases where a global separation of a spacetime into space and time is not possible. Nevertheless, I shall use the term 'global time', but with the understanding that, whenever necessary, it refers to the global temporal aspects of a spacetime manifold.

So far, we have seen only one side of the 'problem of the structure of physical time', viz., the one concerning the use of the notion of time within a spacetime description of physical reality. I shall now turn to the other side which concerns the epistemological status of the structure of physical time.

The birth of non-Euclidean geometries at the beginning of the 19th century not only led to the insight that several alternative geometries had to be taken into consideration in the determination of the structure of physical space, it also led to the recognition that the same applies to the determination of the structure of physical time. But only in the theory of relativity (the special (STR) and the general (GTR) theory) real alternatives for the standard conception of physical time were introduced within physics.

Before the advent of relativity theory (RT) the epistemological problems concerning the structure of physical time were roughly all pervaded by the idea that there was only one possible candidate for the actual structure of time: time as straight line. All other possibilities belonged to the domain of pure speculation.

But RT drastically changed the situation. In the transition from classical to relativistic physics, a number of physically relevant alternatives for the actual structure of physical time presented themselves. For instance, the concept of closed time was no longer purely speculative, but had become a physically significant concept since GTR allowed for spacetimes with closed worldlines; for an observer on such a worldline, time will be closed (circular).

As a result, the following epistemological problem presents itself. Given the fact that various possibilities have to be taken into account for the structure of physical time, and that the actual structure cannot be singled out on a priori grounds, is it possible to turn to experimental data for a solution of the problem of the structure of physical time? At first, an affirmative answer looks plausible. However, the analogous case of the problem of the structure of physical space shows the question to be more complicated. After the discovery of non-Euclidean geometries, it was at first thought, e.g. by Gauss, that a simple measurement would be able to settle the issue whether space was Euclidean or not; in other words, Gauss was convinced of the existence of crucial experiments in that matter. But the situation turned out to be much more complicated. The ensuing discussions concerning the epistemological status of the structure of physical space are too well known to be repeated here. The fact that even today this issue is still a matter of dispute, shows clearly how complicated this problem is. The case of physical time is not simpler than that of physical space. The same kind of problems are encountered and no straightforward answers are to be expected.

In view of these considerations, I intend to expose in this chapter in the first place in what way RT has changed the physicist's conception of time, and, in the second place, how this change affected the epistemological status of the structure of physical time. In approaching these problems I take as my point of departure the fact that it is possible to distinguish between the following kinds of properties of time:

- order properties: 'before', 'after',

- topological properties: continuity of time, boundaries of time, number of dimensions of time, etc.,

- metrical properties: length of time intervals.

In other words, the total structure of time can be conceived to consist of a number of sub-structures: the order, the topological and the metrical structure. However, if spacetime is taken to be the fundamental spatio-temporal entity, then the order, topological and the metrical properties of time have to be derived from respectively the order, the topological and the metrical structure of the spacetime manifold. These three sub-structures are not the only ones that can be distinguished in a spacetime manifold. The full-blown structure of a spacetime manifold can, roughly, be decomposed in the following sub-structures (2):

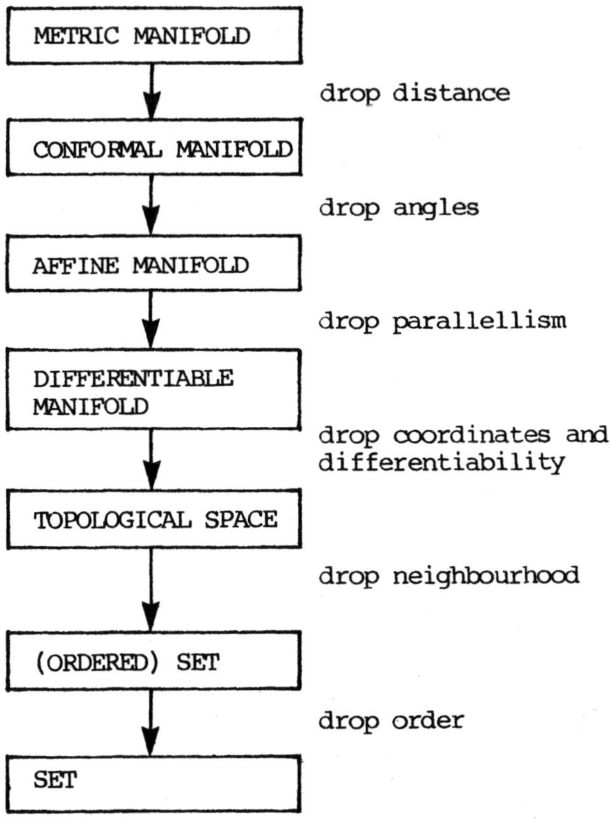

In other words, a metric space M can be construed from a set by defining on this set successively a topology, coordinate systems and differentiability conditions, parallel-displacement, the notion of angle, and finally the notion of distance between two points.

Going from the metric manifold to the underlying set, the structures become increasingly general in nature. This can be seen as follows. All the different types of manifolds (spaces) mentioned in the above scheme can be characterised by a corresponding group of transformations (in the way indicated by Felix Klein in his famous 'Erlanger Programm' [Klein 1921]). But going from the metric manifold to the topological space, these corresponding groups grow larger: the group leaving invariant metric relations is a subgroup of the group leaving invariant angles, etc. The larger the group of transformations, the more general the kind of properties it leaves invariant.

Not all sub-structures necessary for construing the total structure of a spacetime manifold are of equal importance for the study of the structure of physical time. The most important and interesting ones are the order, the topological and the metrical structures. I shall confine myself mainly to these, but occasionally it will also be necessary to consider the conformal structure. It is important to keep in mind that the possibility of a division of the total structure of a spacetime manifold does not imply that all its structural levels are independent of each other (e.g., a given affine manifold cannot be endowed with an arbitrary metric).

The above classifcation of the properties of time proves to be very fruitful. First of all, it makes possible a systematic presentation of the changes in the physicist's conception of time due to RT. We wille see that, whereas STR modified the Newtonian time concept only insofar the order and metrical properties are concerned, GTR also puts into question the topological properties of time. Secondly, it proves to be an excellent guide in finding our way through the numerous epistemological problems, discussed in the literature about the structure of physical time. Each structural level brings forth its own specific kind of epistemological problems.

I will start with discussing the order structure of time (sec. I.2) and then proceed to the topological (sect. I.3) and metrical (sect. I.4) structures. However, before beginning the analysis of the temporal order

relations, the following general observations must be made.
-The reader is reminded that the following presupposition obtains throughout this chapter: spacetime is taken to be the basic spatio-temporal entity; in other words, to speak of space and time is to speak of the spatial and temporal aspects of a spacetime manifold. The paradigmatic instance of a spacetime is a relativistic spacetime which is a four-dimensional differentiable manifold M equipped with a Lorentzian metric (for more details see [Earman 1970, p.274 ff]). The advantage of this point of view is obvious; the relationship between space and time is most easily seen within the spacetime picture. Moreover, certain aspects and problems about space and time can be discussed properly only when space and time are considered not by themselves, but as respectively a three- and one-dimensional submanifold of a four-dimensional spacetime (3). This is especially the case with regard to the global aspects of time.
-In the next chapter I shall introduce the distinction between parameter time and coordinate time. This distinction concerns the role time plays in physical theories; time can be used in two different ways, as a parameter and as a coordinate. This double nature of time with regard to its role in physical theories has, generally speaking, no relevance for the theme of this chapter: the structure of physical time and its epistemological status. An exception has to be made for the topic of the number of dimensions of time; in that case it is useful to take the distinction between parameter and coordinate time into account (see sect. I.3.2).

I.2 The order structure of time

I.2.1 Partial and total order relations.

Of crucial importance for an understanding of the difference in temporal order relations between classical Newtonian time and relativistic Einsteinian time is the distinction between a partial and total order. In a total order, all the elements of a set are ordered into a unique chain (fig. 1a), whereas in a partial order they are ordered in the form of a lattice, with branching points (fig. 1b). This distinction can be expressed more accurately with the help of the formal definition of an order relation. Given a set S, a binary relation R on S is called an order relation if, for all x,y and z, with x,y,z elements from S, the following relations hold:

1) reflexivity : xRx
2) anti-symmetry : xRy and yRx implies x=y
3) transitivity : xRy and yRz implies xRz.

A familiar example of an order relation is the 'smaller than or equal to' relation defined on the reals.

fig. 1a
total order

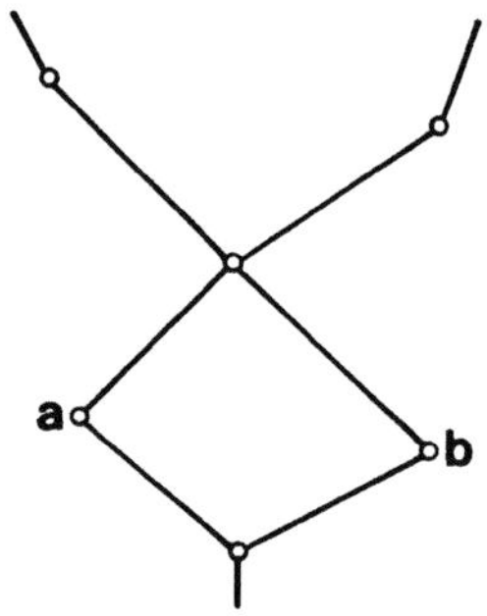

fig. 1b
partial order

A total order relation on the set S is characterised by the fact that for any two elements x,y from S ($x \neq y$) it is true that xRy ór yRx ('ór' in the exclusive sense). On the other hand, in a partially ordered set, there are pairs of elements x,y for which neither xRy nor yRx is true; for example, elements a and b in fig. 1b.

When applying these definitions to temporal order, we have to start from the set S of all events and from the relation R = 'earlier than or simultaneous with' (or, inversely, 'later than or simultaneous with'). This relation, however, is not always a genuine order relation on the set of all events; whether it is an order relation depends upon how physically temporal order is defined. In the case of Newtonian physics, it is not yet an order relation, but it can easily be turned into one, more specifically, into a total order relation. In the case of Einsteinian physics, it is already an order relation, but only a partial one. Both cases will now be considered in more detail.

I.2.2 Newtonian case: total temporal order.

According to Newton, any arbitrary pair of events a and b has an intrinsic temporal order: either a is earlier than b, or the reverse is true, and if neither is the case, then both are simultaneous. This temporal order is absolute, i.e., it is fixed once and for all, and it is the same for all observers. In principle, knowledge or discovery of these intrinsic temporal relations between events is possible by virtue of causal relations. If event a is a possible cause of event b, then a is earlier than b, and vice versa if b is a possible cause of a, then b is earlier than a. If either one is the possible cause of the other, then a and b are simultaneous. (In principle, simultaneity of events is determined by the gravitational force: this force acts at a distance which means that cause and effect are simultaneous.)

Strictly speaking, the Newtonian temporal order relation, R^N, with the above properties is not an order relation at all when it is applied to the set of all events, because the anti-symmetry condition is not met: xR^Ny and yR^Nx does not imply that x and y are the same event but only that x and y are simultaneous. To be a formal order relation, R^N has to be applied not to the set of all events, but to the set consisting of the equivalence classes generated by the equivalence relation 'simultaneous with'. An equivalence relation is reflexive and transitive, but unlike an order relation, symmetric: xRy implies also yRx. Whereas an order relation puts the elements of a set into a (branched) chain, an equivalence relation partitions a set into disjoint subsets, the so-called equivalence classes. In the case of time, the equivalence relation 'simultaneous with' partitions the set of all events into subsets consisting of all simultaneous events. For the sake of convenience, these equivalence classes will be referred to here as 'moments of time'. If R^N is applied to the set of all moments of time, then R^N is an order relation, more particularly a total order relation, since for any two moments of time a and b (a≠b) aR^Nb or bR^Na.

It should be noted that because of the absolute character of Newtonian time, the equivalence relation 'simultaneous with', and the equivalence classes generated by it, have an absolute status. They represent an objective property of physical reality; the partition of the set of events into equivalence classes is the same for all observers. This is no longer the case within relativistic physics.

I.2.3 Einsteinian case: partial temporal order.

In RT the situation with regard to the temporal order of events is different. I will deal separately with the special and the general theory of relativity.

Special theory of relativity. In the Minkowskian spacetime of STR, we can distinguish, due to the indefinite metric, at each point O between the following classes of events (see fig. 2a):

a) events with a timelike distance to O,
b) events with zero distance to O, and
c) events with a spacelike distance to O.

The events with a timelike distance to O fall apart into two disconnected parts, the future and the past, while the events with zero distance to O constitute the lightcone.

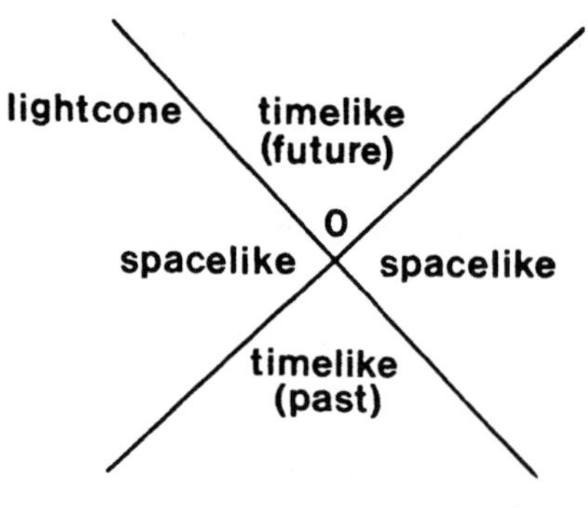

fig. 2a

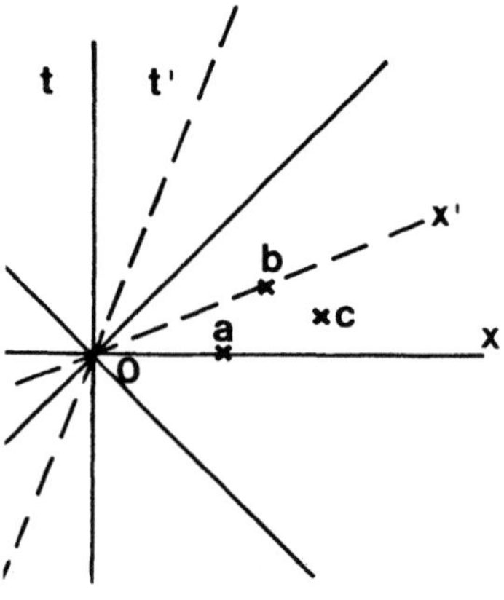

fig. 2b

Only the events within the lightcone of O, i.e., with a timelike distance to O, have an invariant temporal order relatively to O: those in the future lobe of the lightcone are later than O, while those in the past lobe are earlier than O. This is the same for each observer independently of his state of motion at O. But no such invariant temporal order relatively to O exists for the events with a spacelike distance to O. Depending upon the state of motion of the observer, these events are later than, simultaneous with or earlier than O. For example, in fig. 2b, event c is later than O in reference frame (x,t) of observer A, but earlier than O in reference frame (x',t') of observer B (4). Even for one observer, the temporal order of two events with a spacelike distance is not invariant because this temporal order may change if the state of motion of the observer changes. For instance, if observer A of fig. 2b were to change his state of motion at O to the state of motion of observer B, then for observer A the temporal order of the events O and c would be reversed.

Thus, the temporal order of events in STR is only a partial order. Let us take a look at this partial order from the point of view of the simultaneity relation. We have seen that for Newton the relation 'simultaneous with' was an equivalence relation and, moreover, that this relation had an objective status. But in STR simultaneity looses this objective status. Simultaneity is no longer an intrinsic property of the events. Simultaneity of events is not something to be discovered, but to be defined in terms of a specific procedure [Einstein 1954, p. 21 ff]. However, the standard procedure, as given by Einstein, makes simultaneity dependent upon the state of motion of the observer: two observers will regard different sets of events as simultaneous. Therefore the simultaneity relation is not an equivalence relation, as it was for Newton, because it is not transitive: although events a and O are simultaneous for observer A (reference frame (x,t)), and O and b are simultaneous for observer B (reference frame (x',t')), a is not simultaneous with b, neither for observer A nor for observer B. The simultaneity relation is transitive only when applied by one and the same observer and is therefore in that case only an equivalence relation.

Since simultaneity has become observer dependent, the set of all events can not be partitioned, in an objective, observer independent way, into mutually disjoint simultaneity classes. This absence of an observer independent partition of the Minkowskian spacetime into 'simultaneity planes' accounts for the lack of a total temporal order in STR.

Now it is easy to understand why in the case of STR the relation 'earlier than or simultaneous with' is indeed an order relation. This relation satisfies, contrary to the Newtonian temporal order relation, the anti-symmetry condition: if two events a and b are simultaneous for all observers, then this implies that a and b must be the same event. So the anti-symmetry condition obtains.

It is well known that the origin of this lack of a total temporal order within STR is to be found in the causal structure of the world. This is due to the fact that there is an upper limit, viz. the velocity of light in vacuum, to the velocity with which causal influences are propagated. Because of this limit, only the events within the lightcone of O can stand in a causal relation with O (for the events in the future lobe, O is a possible cause, while events in the past lobe are a possible cause of O). These events must have an observer independent temporal order relative to O, otherwise the cause-effect relation between those events would not be preserved.

For events outside the lightcone of O, a causal connection with O is in principle impossible, because of the upper limit in the velocity of propagation of causal influences. For those events it is impossible to determine an intrinsic or objective temporal order relative to O. For this reason, there is only a partial temporal order of events in STR.

The foregoing should not be misunderstood. By selecting a special reference frame, it is possible to construct within STR a total temporal order of the events, in other words to construct a kind of absolute time [Prokhovnik 1967, p. 49 ff]. Contrary to the total temporal order within Newtonian physics, however, such a total temporal order within STR has no objective significance; this point will be discussed extensively later on within the context of the notion of cosmic time function.

<u>General theory of relativity</u>. GTR changes the foregoing picture considerably, since not all the spacetimes

satisfying the field equations of GTR are such that a reasonable definition of temporal order is possible. Consider, for instance, a spacetime that possesses closed worldlines. The temporal order of the events on such worldlines cannot be represented in an adequate way with the help of a binary order relation R. Take an arbitrary pair of events a and b on such a worldline (see fig. 3a); if a is earlier than b, then it is also true that b is earlier than a. Thus for all a and b, aRb implies that bRa, which again implies, if R is an order relation, that a=b, which is clearly absurd. Therefore a binary order relation is unable to represent the cyclic order of events on a closed worldline (5). But also a ternary relation (for instance, 'x lies between y and z') is not powerful enough to adequately describe the cyclic order, since it is impossible to distinguish on the basis of such a relation between the order of events a,b,c and d as given in fig. 3a and fig. 3b; only with the help of a four-place relation, called 'pair-separation', is it possible to distinguish between both orders of a,b,c and d [Newton-Smith 1980, p. 57 ff].

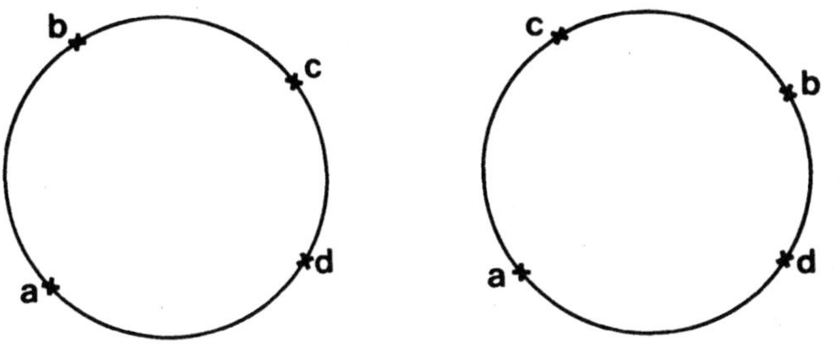

fig. 3a fig. 3b

It follows that within a spacetime with closed worldlines, the use of the relations 'earlier than' and 'lies in between' is, at least globally, no longer meaningful. Only expressions like 'events a and b are pair separated from events c and d' contain in that case any information about the order of events a,b,c and d.

If a spacetime manifold does not admit a binary temporal order relation, then a pre-order relation can very often be meaningfully defined; a pre-order relation is reflexive and transitive, but not anti-symmetric. Kronheimer [1971] has shown that for a large class of spacetimes the following pre-order relation, called 'temporal precedence', obtains: a is temporally precedent to b, whenever there exists a future-directed timelike path from a to b. The reason for selecting timelike paths is obvious; temporal order is intimately connected with causality and points of M which can be connected by timelike paths are points for which a causal relation actually exists or is possible. Temporal precedence, however, is not yet an order relation, since it is not anti-symmetric. But for spacetimes satisfying the chronology condition which states that the spacetime under consideration does not possess closed timelike worldlines, the relation of temporal precedence is also anti- symmetric, and therefore it is in those cases an order relation as well [Hawking 1973, p. 189].

I.2.4 Cosmic time functions.

Although for spacetimes satisfying the chronology condition a reasonable definition of temporal order is possible, this condition is not sufficient to guarantee the existence of 'cosmic time functions': a spacetime M admits a cosmic time function, if there exists a function f from M to R (the reals), such that f increases along every future-directed timelike curve or null curve [Hawking 1968]. Thus, f defines a kind of global time function: it assigns to each event its 'cosmic' time. Hawking has shown that 'stable causality' is a necessary and sufficient condition for the existence of cosmic time functions. Roughly speaking, a spacetime M is said to be stably causal if there are no closed timelike or null curves in any Lorentzian metric which is sufficiently 'near' to the original metric of the manifold M [ibidem].

Spacetimes that allow for the definition of a cosmic time function are of special interest, for it looks as if a total temporal order of the events can be reintroduced in our relativistic models of the universe by using those cosmic time functions, particularly by interpreting those functions in terms of the evolution of the universe. Apparently, each event can be dated in an objective way in terms of the state of the universe at the time of its occurrence. This creates the impression that a total temporal order can be obtained for all events, even for those with a spacelike distance to each other. How can this be reconciled with the fact that these models of the universe are based upon RT, according to which no objective total temporal order of the events is possible?

In the following pages I shall concentrate upon this problem. As will be seen, this seemingly objective total temporal order, based upon the intrinsic cosmic date of each event, is not observer independent and therefore not absolute in the Newtonian sense. In spite of Whitrow's claim [1972, p. 119] to the contrary, there is no intrinsic cosmic date and no clash with the basic principles if RT occurs. To show this I shall first analyse more carefully the notion of cosmic time function. This will lead to a better understanding of the relation between cosmic time functions and the notion 'evolution of the universe'. The conclusion will be that the idea of an objective total temporal order of the events is due to a misunderstanding of the notion 'evolution of the universe'. This notion is very misleading because its meaning is not observer independent.

If cosmic time functions are considered in abstracto, i.e., without relating them to the notion of the evolution of the universe, it is immediately clear that the existence of these functions does not contradict the basic principles of RT. If a spacetime M admits the definition of one cosmic time function f, then infinitely many other cosmic time functions can be defined; f is by no means unique. But this infinity of different cosmic time functions contains members which generate different total temporal orders of the events. RT, however, does not prescribe a choice; it does not specify which is the 'real' cosmic time function with its corresponding 'real' total temporal order. If this were the case, a contradiction with the principles of RT would occur.

The foregoing can be stated in a different way. The choice of a particular cosmic time function introduces a family of simultaneity planes in the spacetime manifold; each simultaneity plane consists of all those events m of M for which f(m) has the same value. But the simultaneity of events defined by f in the neighbourhood of a point of such a simultaneity plane will coincide only for a special observer, in the appropriate state of motion, with the simultaneity of events defined by the Einstein convention for synchronising clocks. Thus to each cosmic time function f, there corresponds a special group of observers for whom the standard definition of simultaneity gives locally the same result as simultaneity defined by the cosmic time function f. Therefore, if it were possible to single out on the basis of objective physical principles, a unique cosmic time function as the 'real' one, then it would also be possible to single out a preferred group of observers. In that case, however, there would clearly be a clash with GTR according to which all observers, in whatever state of motion, are equivalent from a physical point of view.

Since it is impossible to single out, on physical grounds, a class of preferred observers for whom the definition of simultaneity has objective physical significance, it is also impossible to determine which cosmic time function is the 'real' one. Our choice of f is free. This freedom does not, however, imply that the choice has to be made arbitrarily. Not every choice turns out to be convenient or fruitful. For describing the evolution of the universe, certain 'natural' cosmic time functions force themselves upon us.

Indeed the use of cosmic time functions is of special interest in the context of the description of the evolution of the universe: at each moment of cosmic time, the universe can be said to be in a particular state of its evolution. Is an objective total temporal ordering of the events possible in this way? At first sight, it looks indeed as if each event possesses an 'intrinsic cosmic date'. If we assume that the the universe is not static, but evolves in time, and that it does not go through the same sequence of states twice or more often, then each event can be dated by labeling it in terms of the state of the universe at the moment of its occurrence. Thus, it seems possible to establish objectively whether two events are simultaneous or not. The only thing we would have to do is to see whether or not the labels attached to both events refer to the same state of the universe in its evolution.

The situation, however, is not as simple as that. First of all, there would be the need to define the state of motion of the observer who performs the labeling in order to obtain an univocal dating of the events relatively to the evolution of the universe; observers in different states of motion will observe the universe in different ways. Secondly, suppose that the universe is non-homogeneous, then observers at different places will describe the evolution of the universe in different ways. But it is impossible to determine objectively which states correspond to each other and are therefore to be regarded as simultaneous. This shows clearly that, in general, it is very misleading to speak of 'the' evolution of the universe; this notion lacks a definite meaning.

Nevertheless, there is a particular class of universes for which a reasonable definition of 'the' evolution of the universe can be given, albeit in a very special sense. This is the case for those universes for which the so-called cosmological principle is valid. For such universes it is always and everywhere possible to find a special observer, usually referred to as the 'fundamental observer', for whom the universe at large will look homogeneous and isotropic [Misner 1973, sect. 27.3]. The reference frames corresponding to these fundamental observers are called the 'comoving frames'. For these universes a special coordinate system can be introduced, namely the 'comoving synchronous coordinate system' which distinguishes itself from all the other coordinate systems by the fact that the spacelike hyperplanes of constant t coincide with the hyperplanes of homogeneity [op. cit. sect. 27.4]. In this special situation, a particular cosmic time function (or a set of cosmic time functions, but all leading to the same simultaneity hyperplanes) can be selected from all the others by means of a symmetry condition: the simultaneity planes of the cosmic time function must coincide with the hyperplanes of homogeneity. Now, all these fundamental observers have one very nice property in common: because the universe at large looks the same for all of them, they will also see the universe develop in the same way. Relatively to this group of observers, therefore, we can speak properly of 'the' evolution of the universe.

In spite of this, it is not possible to construct an observer independent total temporal order of the events. Labeling all the events relatively to the evolution of the universe seems at first sight to be such an observer independent procedure for constructing a total temporal order; but it is not, since it is also based upon a preferred group of observers, viz., the group of fundamental observers. Although the above mentioned symmetry condition may be a very good reason for selecting this group of observers, the resulting total temporal order of the events has by itself, without reference to this group of observers, no physical meaning whatsoever. So we reach the conclusion that the cosmological principle may suggest a very elegant way of defining a cosmic time function and its corresponding total temporal order, but it cannot be the basis for the construction of a kind of absolute Newtonian time in relativistic models of the universe.

In summary, RT has forced the physicists to reconsider thoroughly the properties of temporal order. But RT has not only led to a careful study of the structure of temporal order, but also of the nature thereof. Indeed, RT disclosed an intimate relationship between temporal and causal order: only those events which can be connected by a causal chain can be ordered temporally in an invariant way. Events for which no causal connection is possible, have no temporal order. Therefore, the temporal order relation 'before' has the same extension as the relation 'is a possible cause of'. But if these two binary relations do have the same extension, why make any difference between them; should we not apply Occam's famous rasor? Would a reduction of temporal order to causal order (or vice versa) not be preferable?

I.2.5 Causal theories of time.

In recent years, the problem of reducing temporal order to causal order has received considerable attention; it is usually referred to as 'the problem of the causal theory of time' (6). All causal theories of time are reduction theories: in some way or another they try to reduce temporal relations to causal relations. Different versions have been proposed. In the first place, causal theories of time differ with respect to the choice of the type of causal relations to which temporal relations are to be reduced. For instance, Reichenbach tried to reduce the temporal order relation 'before' to the asymmetric causal relation 'is the cause of' ('asymmetric' here means that if a is the cause of b, then b cannot be the cause of a) [Reichenbach 1957, p. 135 ff]. This required a distinction between cause and effect without a recourse to temporal relations. Reichenbach thought this to be possible for "causality establishes not a symmetrical but an asymmetrical relation between events" [op. cit. p. 136]. More in particular, he proposed the following 'a-temporal' characterisation of cause and effect [ibidem]:

> If E^1 is the cause of E^2, then a small variation (a mark) in E^1 is associated with a small variation in E^2, whereas small variations in E^2 are not associated with variations in E^1.

However, Grünbaum [1973, ch. 7] has shown convincingly that Reichenbach's proposal for reducing temporal to causal order is not acceptable and instead has made the suggestion to define temporal betweenness in terms of the symmetrical causal relation 'is causally connectible with'. But with regard to his own version, he had to admit that, so far, every attempt to explicitly define this primitive symmetrical causal relation without using any temporal concepts had failed.

In the second place, causal theories of time differ with respect to the extent of the reduction. In weak versions, the reduction is said to pertain to temporal order relations only, whereas in strong versions, all temporal relations (order, topological and metrical) are claimed to be reducible to causal relations.

The counterpart of a causal theory of time in the spacetime approach is a causal theory of the structure of spacetime which means that spatial as well as temporal relations between events must be reduced to the causal structure of the world. Winnie [1977] has treated this problem thoroughly in his article "The causal theory of space-time". He claims to have shown that the entire structure (order, topological and metrical) of Minkowski spacetime can be derived from the causal structure of the special theory of relativity:

> while special relativity has been cited by defenders of a causal theory of time as providing fresh support for the causal theory of <u>temporal order</u>, these same causal theorists have <u>not seen that</u> the causal resources of this theory are so

> powerful as to yield in addition a purely causal account of temporal metric, spatial geometry, and the topology and metric geometry of Minkowski space-time.

This amounts to an extremely strong version of the causal theory of spacetime (7).

Instead of entering into a detailed discussion and evaluation of these different kinds of causal theories of time (spacetime), I shall restrict myself to making a few remarks concerning the usefulness of these attempts to reduce temporal relations to causal ones.

In spite of the sometimes remarkable success of these reduction theories, the question remains whether it is at all necessary and (or) useful to reduce spatial and temporal relations to causal ones. Among others, Sklar [1977, p. 333 ff] has raised serious doubts about the need for such theories, and argues that causal theories of time (spacetime) are rather implausible. Instead, he is more in favour of a reduction in the reversed way, namely a 'spatiotemporal theory of causation'. One of the reasons he adduces for the "superiority of a spatiotemporal theory of causation to a causal theory of time" is that, in the latter, "actual temporal relations are associated .....with only the possibility of the appropriate causal relations" [op. cit. p. 342; see also Denbigh 1981, p. 32].

Another argument against the usefulness of causal theories of time stems from a consideration of the general nature of reduction theories. These have (or should have) at least two characteristics. In the first place, a reduction theory aims at translating all terms and concepts from a less fundamental level into those of a more fundamental one; it always involves a hierarchy of levels. Secondly, such theories translate vague and not well understood concepts into more clearly defined and better understood concepts. A reduction always involves an increase in the understanding and the clarity of the basic concepts. But neither condition is satisfied by the causal theories of time.It is not self-evident at all that causal relations are more fundamental than spatial and temporal ones. Furthermore, it is highly dubious that the nature of causal relations is better understood than the nature of temporal and spatial relations.

The foregoing remarks are not intended to show that causal theories of time are superfluous; on the contrary, they have contributed very much to a better understanding of the relation between time and causality. My intention was to show that insofar as these theories aim at a complete reduction of temporal to causal relations, they do not lead to a better understanding, for the reasons given above, of the nature of physical time.

Besides necessitating a critique of longstanding ideas about the temporal order of events, RT has also caused a change in the physicist's conception of the topological structure of physical time, primarily by revealing the complex nature of the global structure of time, but also by enabling the physicist to restate old, speculative ideas about time, such as the idea of cyclic time, within the context of a physical discourse.

I.3 The topological structure of time

I.3.1 Time and the topology of spacetime.

Various apparently unrelated problems concerning the concept of time are nevertheless mathematically of the same kind. Questions as diverse as "How many dimensions does time have?", "Is time continuous or discrete?",

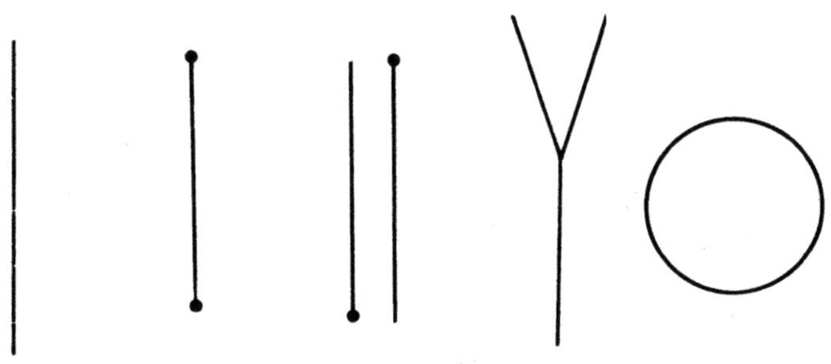

4a 4b 4c 4d 4e

fig. 4

"Does time have the structure of an Euclidean line (fig. 4a), of a closed line segment (fig. 4b: time with beginning and end), of a semi-line (fig. 4c)?", "Is time many-fingered (fig. 4d), or cyclic (fig. 4e)?" and "Is time orientable?", all concern the problem of the topological structure of time.

Whereas STR changed the physicist's ideas about temporal order and, as will appear in sect. I.4, also about the measurement of time intervals (the metric of time), it did not modify the ideas about the topological structure of time. Time, within STR, has the same topological properties as absolute Newtonian time, viz., those of the Euclidean line (fig. 4a). It was not until GTR was developed that the 'field of possibilities' for the topological structure of physical time became enlarged.

Since it is not possible to treat here in extenso all the topological properties of time, I shall limit myself to two central topics: first, the problem of the continuity versus discreteness of time, and the related issue of the number of dimensions of time, and secondly, the notion of temporal orientability. The first topic concerns the local structure of time while the second deals with a global aspect of time. The importance of the first problem lies in the fact that the notion of 'discrete time' is sometimes proposed as a possible solution to fundamental problems encountered in microscopic physics; but in most cases, the notion of discrete time itself remains rather obscure. The second topic deserves our attention because temporal orientability is related to the issue of time asymmetry which will be discussed in chapters III and IV.

A discussion of these topological properties of time first requires some general observations about i) the topological structure of spacetimes, and ii) the topological structure of time within a spacetime approach.

i) Topology of spacetimes. Topology treats of the 'neighbourhood'-relations between the points of a space. Loosely, it can be described as 'rubber sheet geometry', because it studies only those properties of geometrical figures that remain the same no matter how the rubber sheet on which these figures are drawn is stretched (without tearing or gluing). Formally, topology deals with those properties that are left invariant under homeomorphisms: one-to-one bicontinuous mappings. These homeomorphisms do not change the neighbourhood relations between the points of a space. A set is turned into a topological space by specifying a family of open subsets T of S, which is closed under union and finite intersection. In the case of metric spaces, the family of open sets defining the topology is usually given with the help of the metric. This metric of the space M induces in a natural way a topology on M, of which the open sets correspond to the ε-balls around the points of M (8).

For spacetime manifolds, however, the relation between the spacetime metric and the spacetime topology is more complicated. The topology of a spacetime manifold cannot be given with the help of its metric, because the spacetime metric is not positive definite. The open sets induced by the spacetime metric do not specify a topology. Indeed, all the points of a spacetime manifold that correspond to the path of a light ray have zero distance with respect to each other, so that any ε-ball around any of these points contains all the points of the light path; but, intuitively, it is clear that not all these points are 'near' to each other.

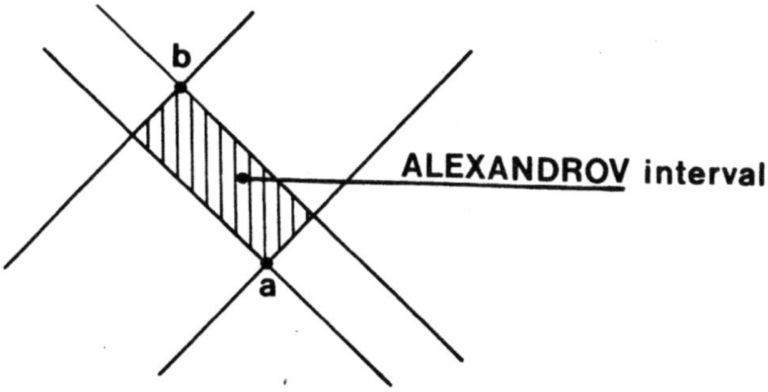

fig. 5

However, the topology of a spacetime manifold can be defined rather simply with the help of the temporal order relation [Kronheimer 1971, Winnie 1977]. Essentially, this works as follows. The basis of open sets of the topology is defined by the Alexandrov-intervals. The Alexandrov-interval of two events a and b consists of all events chronologically between a and b, i.e., all events inside the intersection of the lightcones of a and b (see fig. 5). The topology thus defined is called the Alexandrov topology which coincides with the spacetime manifold topology in those cases where 'strong causality' holds (9).

Yet, in spite of the intimate relationship between the topological and the order structure of a spacetime manifold, it remains that both are of a different nature. Not all topological properties can be expressed in terms of the order structure. Therefore it is important to clearly distinguish between both kinds of properties (10).

ii) Topology of time. Given a spacetime M, how can the topology of time be derived from the topology of spacetime? I will treat this problem both from a local and a global point of view.

- The topology of time from a local point of view. Locally, the situation is rather simple, since locally any spacetime manifold looks everywhere the same, viz., like a flat Minkowski spacetime. The latter is the direct product of E^3 (Euclidean space) and E^1 (time). This means that locally, time has everywhere the same topological properties on the spacetime manifold. In fact, there is but one topological property which completely determines the local topological structure of the manifold: the number of dimensions [Thirring 1977, p. 12]. Within the standard picture of time as a straight line, this number of dimensions is equal to one, since time is regarded to be homeomorphic to the one-dimensional Euclidean line E.

- The topology of time from a global point of view. In an article entitled "How to talk about the topology of time", Earman [1977b] has discussed this difficult problem in detail, especially with regard to the question whether time is open or closed. He attacks the problem in two different ways to which he refers as the injection and the projection approach to the problem of the topology of time. The idea behind the injection approach is to "relativize talk about the topology of time to a given space point" [op. cit. p. 213]. In that case, the topological properties of time are derived from the topological properties of a given set of worldlines on M (defined by a timelike vector field V on M). A characteristic feature of this approach is that time can be open at one point of space and closed at another, depending on whether the worldlines through these points are open or closed. Moreover, the topological properties of time may depend upon the choice of the set of worldlines on M (or the corresponding choice of V). Earman gives the example of the Gödel universe. On this spacetime, two different sets of worldlines can be defined, one leading to the conclusion that time is closed at each point of space, the other to the conclusion that time is open at each point of space. Thus, it is not the spacetime manifold itself that determines completely the topological properties of time at a certain point of space, since the determination of these topological properties depends also upon the given set of worldlines, at least within the injection approach.

In the projection approach it is assumed that M admits a partition into a family of 'time slices' Y (11). In that case, the projection topology of the factor-space M/Y can be regarded as the topology of time. This approach, however, has the disadvantage that the assumption about the partitionability of M into time slices is very strong and therefore severely restricts the applicability of the projection method. For instance, it cannot be applied to the Gödel universe. Moreover, it leads to rather awkward conclusions such as the possibility of a closed time for a Minkowski spacetime [Earman 1970, p. 215].

Neither the injection nor the projection approach allows for an adequate treatment of the question whether time is open or closed. It seems to be doubtful whether it is at all possible to develop a general procedure for determining unambiguously the topological structure of time for a given spacetime manifold. The following complication occurs. Let us suppose that, according to a given procedure, a spacetime M has a non-standard topological structure for time, e.g., that time is closed. In that case, this non-standard temporal feature of M can often be eliminated by going to a so-called 'covering spacetime', M', which is observationally indistinguishable from M, but which does not exhibit this non-standard temporal feature (12).

An example of a transition from a spacetime M with closed time to a covering spacetime with the standard topological structure for time is given schematically in fig. 6 (13). For an observer in spacetime A, time is closed, but for an observer in spacetime B, which is observationally indistinguishable from spacetime A, time is open. Thus it should be kept in mind that, as Earman [1977b, p. 225] remarks: "non-standard time structures (e.g. branching and circular times) have a non-intrinsic character in that under very general conditions they can be interpreted away in favor of standard time structures." There are exceptions; for example, the closed timelike curves in a Gödel spacetime; they cannot be eliminated by going to a covering spacetime (14).

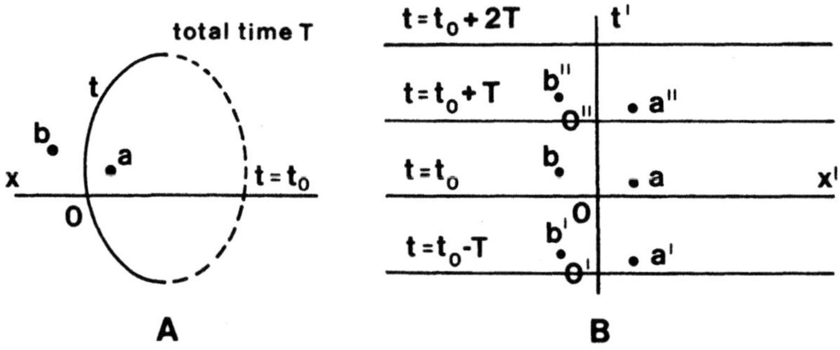

fig. 6

The upshot of the foregoing is that speaking about the topology of time, and particularly, about the global properties of time, is at least hazardous within the spacetime approach; it is "harder to talk meaningfully about the topology of space and time than has been generally realised" [Earman 1977b, p. 225].

I shall now turn to a discussion of the local topological structure of time.

I.3.2 Discrete versus continuous time.

Physical time in the standard picture is a one-dimensional continuous manifold. This is the case for both Newtonian absolute time and time in RT (in RT time is considered to be a one-dimensional submanifold of the four-dimensional spacetime manifold). These two properties of time, its continuity and one-dimensionality, are usually taken for granted; whenever their self-evidence is put into question, then discrete time or multi-dimensional time are mentioned as alternatives to the standard picture.

What does it mean for time to be continuous and one-dimensional? A proper treatment of this question makes it necessary to recall the definitions of continuity and of the number of dimensions of a space.

Continuity. The fact that the elements of a set constitute a continuum is related to the type of ordering of the elements of the set. Given a set S with a total order relation R, it is possible to distinguish at least three different orders: a discrete, a dense and a continuous order.

- Discrete order. Let S be the set of natural numbers and R the usual arithmetical order relation. This type of order is called discrete. The number of elements between any two elements of the set is finite; it is characteristic for this type of order that the notion of next element is well defined.

- Dense order. Let S be the set of rational numbers and R again the arithmetical order. This is a dense order; the number of elements between any two elements of S is not finite but denumerably infinite, and the notion of next element is meaningless.

- Continuous order. Let S be the reals and R as above. In this case the number of elements between any two elements of S is superdenumerably infinite. Here too, the notion of next element is meaningless.

Intuitively, the most conspicuous property of a continuum is the fact that it does not have any 'gaps' [Weyl 1963, p. 39-40]. For the reals, this is reflected in the fact that they are Cauchy complete (15). Cauchy completeness formalises this absence of gaps in a continuum in the following way: in a continuous set, any converging sequence of elements of S has a limit which is an element of the set itself. It can be shown easily that the dense order of the rationals does not meet this requirement and that they are therefore Cauchy incomplete. Consider all the rationals smaller than the square root of two; clearly the sequence composed of these does converge, but it does not converge to an element of the set of rationals itself. The set containing both the rationals and all these limit numbers is the set of the reals; this set has the very nice property that any converging sequence of real numbers has a limit which is itself a real number. There are no gaps in the reals.

Returning to time, let S be the set of instants of time and R the 'earlier than or simultaneous with' relation. Again, three types of temporal order can be distinguished, viz., a discrete, a dense and a continuous temporal order. For discrete time, the number of instants in a time interval is always finite and the notion of next instant is well defined. For dense, respectively, continuous time, the number of instants in a time interval is denumerably, respectively, superdenumerably infinite; for both the notion of next instant is meaningless. Dense and continuous time are infinitely divisible in the sense that for any time interval there exists an instant which lies exactly halfway, given the standard metrics; this is not the case for discrete time.

It should be remarked that the distinction between discrete, dense and continuous time is based upon the order type of the instants of time and only involves set-theoretic concepts. It is not necessary that a topology is given on S. Therefore, continuity (in the sense of completeness), denseness and discreteness, as defined above, are not strictly topological concepts. Nevertheless, as will be seen shortly, there is an intimate relationship between a discrete order type of a set and a discrete topology.

The number of dimensions of a space. The number of dimensions of a space is a genuine topological property. I will make use of the definition of Menger because it corresponds closely to our intuitive view of the number of dimensions of a space (16). It formalises the fact that, for instance, a three-dimensional room is bounded by two-dimensional walls. Menger's definition runs as follows:

i) The empty set has dimension -1

ii) The dimension of a space is the least integer n for which every point has arbitrarily small neighbourhoods whose boundaries have dimension less than n.

According to this definition, the Euclidean line E with the usual topology is one-dimensional: dim(E)=1. An interesting consequence of this definition is that both discrete and dense time are zero-dimensional, since the only possible topology on a set with a finite or denumerably infinite number of elements is the discrete topology and, for any set with a discrete topology, the number of dimensions is zero [Hurewicz 1941, p. 10-11]. Thus, discrete time is zero-dimensional. On the other hand, continuous time is not necessarily one- or more-dimensional, since continuous time with the discrete topology is zero-dimensional.

Because the number of dimensions of a space is a topological property, the one-dimensionality of time cannot be inferred directly from the temporal order properties, in particular not from the fact that the instants of time are linearly ordered. Indeed, it is possible to construe a one-to-one correspondence between the points of a plane and the points of a line [Kline 1972, p. 997]. Thus, the points of a plane can be given a total order, and the same applies for the points of n-dimensional space. Likewise, the instants of two- or n-dimensional time can be arranged in a linear, total order. But of course, the open sets defined with the help of this linear order do not form a basis for the topology of two- or n-dimensional time.

Thus the linear order of the instants of time does not lead immediately to the conclusion that time is one-dimensional; at least two additional assumptions are necessary to arrive at this conclusion. First, the instants of time must constitute a continuum under the temporal order relation 'earlier than or simultaneous with'. Secondly, since this still leaves open the possibility for zero- and n-dimensional time, the open sets defined by this order relation must form a basis for the topological structure of time. Without these two assumptions, non-standard times like zero-dimensional discrete time or multi-dimensional continuous time cannot be ruled out. Since both these non-standard time conceptions are occasionally put forward by physicists, it is worthwhile to consider them in more detail.

Zero-dimensional discrete time. In the standard conception of discrete time, the process of dividing a time interval is supposed to have a 'natural end'; in carrying out the process of division, a situation is reached wherein it is no longer possible to divide the remaining time interval in still smaller parts. Time is conceived as built up from time-atoms, also called time-quanta or chronons. Within this picture of discrete time, the time-atoms are assigned a fixed duration. Physicists have tried to estimate the length of these time-atoms on theoretical grounds; values in the order of 10^{-23} or 10^{-24} are often mentioned (17).

I will now try to make this picture of discrete time more precise in order to see whether it is coherent. The number of time-atoms in a time interval of discrete time is always finite; if not, time would be dense or continuous. This implies, as we have observed already, that discrete time is zero-dimensional. Nevertheless, this zero-dimensional discrete time is usually depicted as built up from one-dimensional time-atoms which are assumed to be homeomorphic to the standard closed interval. Grünbaum [1973, p. 460] calls this kind of space 'singly extended discrete space'. This picture seems to save the one-dimensionality of time by not ascribing it to time as a whole but to the internal constitution of the time-atoms. The question is whether this can be done coherently. The answer will be negative, because this picture of discrete time leads either to two unrelated temporal order relations, one inside the time-atoms, and one for the time-atoms as a whole; or a tacit appeal has to be made to a kind of continuous 'background time' in which the time-atoms are embedded.

In short, the situation is as follows. Let R^e be the 'external' order relation which arranges the time-atoms into a chain, and let $R^{i,n}$ be the 'internal' order relation which orders the 'instants' of the n-th time-atom. These two temporal order relations are in principle completely independent from each other and they order two different types of instants of time. For discrete time as a whole, the time-atoms are the instants which are ordered by the order relation R^e, whereas, inside the time-atoms, there is a second type of instants which form a bounded continuum and are ordered by the $R^{i,n}$. Thus the usual picture of discrete time involves two different time concepts of which one is discrete (the order of the time-atoms) and the other continuous (the order inside the time-atoms).

A way out of this unsatisfactory situation seems to be offered by the additional embedding of all the time-atoms in a one-dimensional continuous background time. This corresponds to the intuitive picture of discrete time, in which time-atoms are supposed to lie 'next' to each other, or to 'touch' each other [Whitrow 1972, p. 141]:

> Many people have difficulty in imagining time to be "atomic" in structure because they believe this would imply the existence of temporal gaps which must themselves be part of time in contradiction of the hypothesis. On the contrary, however, the "atomicity" of time refers only to the indivisibility of the chronon. In principle, chronons could be imagined as being like a row of pebbles which touch each other, so that there would be no gaps between them.

This way of looking at discrete time solves the problem of the two independent time concepts, since the embedding of all the time-atoms in a continuous background space yields one unique temporal order relation which encapsulates both R^e and the $R^{1,n}$. This overall temporal order relation corresponds to the total ordering of the points of the one-dimensional continuous background space. But this concept of time can hardly be called discrete, since the original discrete time-atoms have become closed intervals (whether or not lying next to each other or touching each other) of this one-dimensional continuous background time which functions as a kind of 'supertime'. Moreover, the discrete topology of time is lost, since the time-atoms are not closed and open time intervals.

Clearly then, the association of the notion of one-dimensional continuous time-atoms with discrete time is problematic. The same applies for the notion 'minimal duration' which stands for a minimum or lower bound (unequal zero) in the accuracy of time measurements. Whether or not such minimal duration actually exists is rather a controversial matter (18). It is, however, interesting that very often the existence of a minimal duration is regarded as a consequence of the discreteness of time: the exact location of the moment at which an event takes place within a time-atom would be impossible. But the reverse is also thought to be true; the existence of a minimal duration is considered as conclusive evidence for the discreteness or quantisation of time [Flint 1948, Abramenko 1958].

There is, however, no logical relationship between the notions of discrete time and of minimal duration. A lower bound in the accuracy of time measurements is compatible with both discrete and continuous time. The quantisation or discreteness of time does not imply the existence of a minimal duration anymore than the quantisation of, e.g., the angular momentum in quantum mechanics implies the existence of a lower bound in the accuracy with which the latter can be measured. There is no reason to assume that, in principle, the outcome of a time measurement in case of discrete time could not be exact; only for discrete time does the set of all possible outcomes of a time measurement constitute a discrete set (19).

Clearly, the fact that the standard picture of discrete time, as built up from one-dimensional time-atoms, is obscure and incoherent does not imply that this notion itself can be disregarded completely; there is no reason to believe that any interpretation of discrete time is necessarily incoherent (especially not if the notion of a one-dimensional time-atom is dropped). Therefore it appears that discrete time stays a viable alternative for continuous time.

Multi-dimensional time. There have been only a few attempts to use multi-dimensional time in physical theories and, so far, these theories have not been very fruitful. For instance, unified field theories occasionally make use of a five- instead of a four-dimensional spacetime manifold [Graves 1971, Ch. 15]. At first sight, such theories open the way for introducing a two-dimensional time. Usually, however, the interpretation of the fifth coordinate as a second time coordinate is based solely upon the fact that this fifth coordinate has the same sign in the spacetime metric as the original time coordinate. It is highly questionable whether this formal analogy between the fifth coordinate and the time coordinate provides sufficient reason to speak of a second time coordinate, and thus, of two-dimensional time.

A special problem with regard to multi-dimensional time is that it seems necessary to introduce a flow of time and the corresponding distinction between past, present and future for each dimension of time. In this respect, the distinction between parameter time and coordinate time (see Ch. II) is of special interest, for it offers a chance to avoid a multiplicity of time flows and of distinctions between past, present and future in the case of multi-dimensional time. In chapter II it will be argued that the notions of past, present and future pertain to parameter time and not to coordinate time, because these notions are related to the flow of time. Coordinate time does not comprise those aspects of time which are related to temporal becoming, and therefore, the notions 'past', 'present' and 'future' have no meaning with regard to coordinate time. From this point of view, the multi-dimensionality of coordinate time does not involve the existence of different time flows, contrary to the multi-dimensionality of parameter time. Therefore, multi-dimensional coordinate time appears to be more readily acceptable than multi-dimensional parameter time.

Nevertheless, the introduction of discrete or multi-dimensional time into physics would, without any doubt, raise many conceptual problems. In spite of these, both conceptions of time constitute real alternatives for the standard conception of time. Therefore, physicists have often tried to prove that physical time is one-dimensional or that spacetime is four-dimensional [Weyl 1952, p. 283, Weyl 1963, p. 136, Dorling 1970, Penney 1965, Lanczos 1962, Scheurer 1979, p. 240]. But, without exception, all these attempts take at least for granted the fact that space, time or spacetime is continuous. This assumption is of particular interest from an epistemological point of view; how can it be justified?

Let us begin by considering the continuity of physical time from the point of view of empirical data. Certainly, the continuity of time cannot be derived directly from these data. In his discussion of the 'empirical significance of continuity postulates', Newton-Smith [1980, p. 121 ff] reaches the conclusion that all relevant data compatible with Newton's mechanics which is based upon continuous time, are equally compatible with 'Notwen's' mechanics which is based upon dense time. According to Newton-Smith this is a clear case of underdetermination of theories by experimental facts. In other words, it would be impossible to design a crucial experiment which would enable us to make a decision between Newton's and Notwen's mechanics, based respectively upon continuous and dense time. In this respect, it is instructive to note that in recent work on the foundations of the differentiable and topological structure of the spacetime manifold, the continuity of spacetime is either explicitly introduced by way of an axiom, or implicitly assumed [Winnie 1977, p. 140, Woodhouse

1973, p. 496].

As for the choice between discrete and continuous time, the situation appears to be the same as for dense and continuous time. I have already pointed out that, contrary to what is very often assumed, the existence of a minimal duration is compatible with both discrete and continuous time. Thus, even if it could be shown conclusively that a minimal duration exists - it is highly questionable that this could be done - then the question whether time is continuous or discrete would still be open.

Assuming that a decision cannot be reached on the basis of empirical data, one could still try to justify to choice for continuous time by pointing to the fact that giving up the standard picture of time creates a great number of new conceptual problems. But a fair comparison of the conceptual problems involved in the different time concepts and the corresponding physics, would be possible only if the alternatives were worked out with the same degree of sophistication as the standard approach, assuming that this can be done. This is not the case. These alternatives have never been developed so extensively as to become serious competitors for standard physics based upon continuous time. But even if they had been developed, then it would still be necessary, in order to evaluate these different approaches, to compare and evaluate the various conceptual problems involved; this would be a difficult task [Laudan 1977, Ch. 2]. It is highly dubious whether the choice for continuous time can be justified convincingly in this manner.

In view of this situation, a hard-boiled conventionalist might claim that, since all alternatives are equally compatible with the available data, the decision to use continuous or discrete time has to be reached by a convention, and that therefore the problem is really a pseudo-problem. However, I think most physicists would judge the situation otherwise; for them it does matter whether time is continuous or discrete. There can be no doubt that the hypothesis of the continuity and one-dimensionality of time was extremely fruitful during the last three centuries: almost all basic laws of physics are given in terms of differential equations. In general, there is no practical need for the physicists to put this hypothesis into doubt. The interesting question, which must be left unanswered here, pertains to the conditions under which physicists would be prepared to reconsider or to give up the standard picture of one-dimensional continuous time and would start working on basis of one of the alternatives.

I.3.3 Temporal orientability.

I shall first have a short look at temporal orientability within the space and time approach and then go to the definition of temporal orientability of a spacetime manifold (20). In principle, nothing new is added in the latter case; only the details are more complicated.

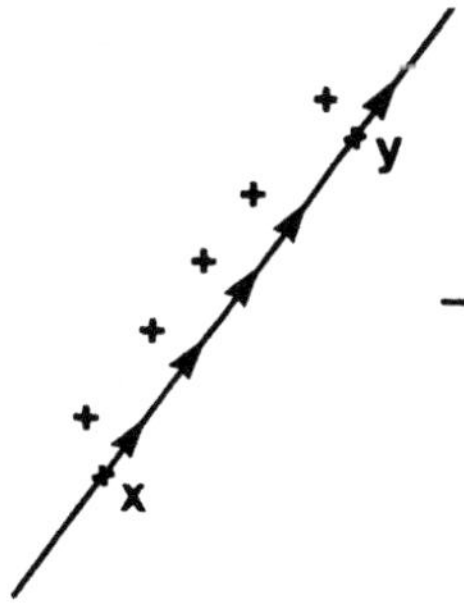

fig. 7a

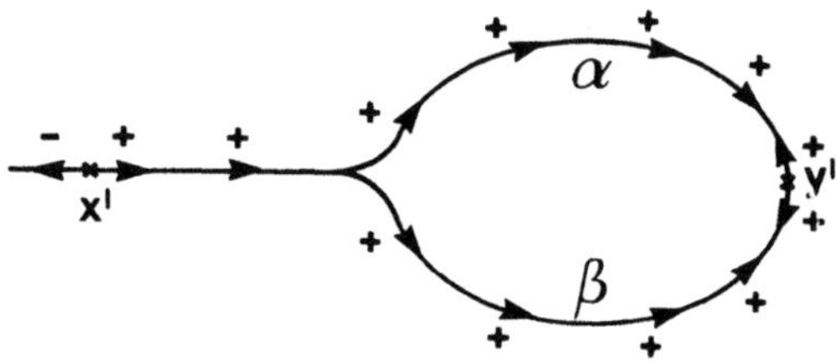

fig. 7b

In the neighbourhood of each moment of time, i.e., locally, a future and past orientation of time, denoted respectively by '+' and by '-' orientation, can be defined. The question is whether this can also be done for time as a whole, whether a global distinction between past and future orientation of time can be defined coherently. This is not always possible. Compare with each other figures 7a and 7b, representing respectively linear time and some rather 'pathological' time which is a weird mixture of linear and circular time. At the points x and x', we choose a future orientation for time (indicated by the + arrow). Starting from the future orientation at these points, the future orientation at the points y and y' is defined in the following way: transport a vector pointing in the + orientation from x (x') to y (y') so that during transport the vector stays parallel to the line representing time. The future orientation at y (y') is then given by the orientation of the transported vector at y (y'). For the time of fig. 7a, this definition leads to a globally coherent distinction between two opposite time orientations. This is not so for the time of fig. 7b. Going along route I to point y' and along route II, does not lead to the same result for the future orientation at y'. Starting from x' with a future pointing vector, we end up again at x', after having traversed the whole loop, with a vector pointing in the past direction. Therefore, a global distinction between past and future orientation of time is not possible in the situation of fig. 7b; such a time is called non orientable. Time is called orientable if, as in fig. 7a, a global distinction between past and future orientation is possible.

Temporal orientability of a spacetime manifold is defined in the same vein. Again we make use of transporting timelike vectors along curves. Take an arbitrary point x of M; at x, all timelike vectors fall apart into two classes, those pointing in the past and those pointing in the future orientation of time. Next, take any closed curve through x, and choose a future (or past) orientated vector at x. Transport this vector in a continuous way along this curve keeping it timelike. After transporting it in this way along the closed curve, two situations can occur: the original vector and the transported one have the same orientation at x (i.e. both lie within the future or past lobe at x), or they have opposite orientations. In the former case, the curve is said to be orientation preserving; in the latter, it is not. Now, a spacetime M is called temporally orientable if and only if for all x every closed curve through x is orientation preserving. By selecting one of the two orientations, a temporally orientable manifold becomes a temporally orientated manifold [Choquet-Bruhat 1977, p. 201]. Spatial orientability of a spacetime manifold is defined analogously.

If a spacetime M is temporally orientable, it does not have to be spatially orientable; if it is both spatially and temporally orientable, then it is also orientable as a whole. Generally, if a spacetime is orientable in two of the three ways mentioned (temporally, spatially, as a whole), then the third one also follows.

Temporal orientability is not a purely topological property of the spacetime manifold M; in certain cases it is related to the metric of M. However, it can be shown that every spacetime which is simply connected is both space and time orientable, independently of its metric [Geroch 1979].

Whereas the details of the definition of temporal orientability are well understood, its physical significance is rather obscure. The problem is: in what physical sense does a temporally orientable spacetime differ from a non orientable one, and can temporally non orientable spacetimes be ruled out, on physical grounds, as possible models for our universe. This is a complicated matter, especially because it concerns one of the global aspects of spacetimes.

A discussion of the empirical significance of the temporal (non) orientability of a spacetime manifold must take into consideration the well known fact that, for any temporally non orientable spacetime, a corresponding covering spacetime manifold can be found that is temporally orientable [Sklar 1977, p. 315]. The problem can therefore be put in the following way: is it possible to distinguish empirically between a temporally non orientable spacetime M and its temporally orientable covering spacetime M'? If not, then it would always be possible to eliminate temporally non orientable spacetimes in favour of their corresponding temporally orientable covering spacetimes.

A strong argument for the physical indistinguishability of a temporally non orientable spacetime and the correponding temporally orientable spacetime is due to Geroch and Horowitz [in Hawking 1979]; in short, it runs as follows. Locally, both manifolds look exactly the same, although globally they are different, since the "only effect of taking the universal covering spacetime is to possibly produce several copies of each local region in the original spacetime" [Hawking 1979, p. 231]. Now, in view of the fact that all fundamental laws of physics are local laws, it is impossible to distinguish on the basis of these laws alone between both types of spacetimes, since they differ only globally. The conclusion would therefore be that, as far as the laws of physics are concerned, the temporal (non) orientability is not a significant property.

But although there is no direct link between the local laws of physics and the question of the orientability of spacetimes, there nevertheless is an indirect connection between both, which concerns the time reversal invariance of physical laws (21). In a temporally non orientable spacetime, the laws of physics have to be time reversal invariant. It is easy to see why this has to be the case. Given that some laws of physics are not time reversal invariant, it follows that processes governed by these laws take place relatively to one orientation of time only, regardless of where and when they take place. Therefore, time reversal non invariance of physical laws implicitly presupposes that it makes sense to speak of two opposite time orientations for the whole manifold, i.e. globally, and thus it presupposes temporal orientability. On the other hand, time reversal invariance of physical laws is compatible with both temporally orientable and non orientable spacetimes.

Thus the existence of time reversal non invariant laws implies that temporally non orientable spacetimes can be eliminated as possible models for our actual universe. For the moment, it is assumed that at least for one kind of processes, the laws governing them are not time reversal invariant, viz., the famous T*-violation in weak interaction processes [Christenson 1964]. Nevertheless, this T*-violation does not allow the conclusion that the spacetime models for our actual universe must be temporally orientable since the assumption that T*-violation occurs in those processes is based upon the validity of the CPT* theorem and the observed violation of CP (22). Therefore the T*-violation in those processes is coupled with CP-violation; this implies that [Geroch 1979, p. 229]:

> the strongest conclusion to be drawn from such arguments, using the presently observed symmetry violations in elementary particle physics, is that our spacetime must be total-orientable. One cannot conclude from this, for example, that our spacetime must be separately time- and space-orientable.

Therefore, the question whether spacetime is temporally orientable or not is still open. There is no convincing argument that it should be temporally orientable, as it is usually assumed to be. In chapter III, I will show that the existence of time asymmetry in our universe does not imply that spacetime is temporally orientable.

Reviewing the foregoing, it is clear that the determination of the topological structure of physical time is not a simple matter. We have seen that it is not necessary to elevate the standard picture of physical time to an unassailable dogma, since the available data are not powerful enough to determine completely and unambiguously the structure of physical time. This conclusion should not be misunderstood; in my opinion, it does not show a defect of physical theories, on the contrary, it gives room to the scientific imagination.

I.4 The metrical structure of time

I.4.1 Clocks in Newtonian and Einsteinian physics.

The complete description of the structure of physical time requires more than the order and topological structures, for these do not capture the notion of length of time intervals. This aspect of time, which concerns duration or the amount of time elapsed between two events, is formalised in the metrical structure of time (23). As soon as a metric is ascribed to time, it makes sense to speak about the length of a time interval, which can be measured by a clock, and about two time intervals having the same length.

Within classical Newtonian physics, the metric of time is completely independent from the metric of space. Both space and time have their own metric. The metric of time is a measure for the flow of absolute time which in Newton's famous words "flows equably without relation to anything external." Two observers using identical, ideal clocks, i.e., clocks which, without being disturbed, correctly measure the flow of absolute time, will always agree on the amount of time they measure for any time interval, regardless of, for instance, their state of motion. Thus the time measured between events A and B along curve I by an ideal clock is equal to the time measured along curve II by an ideal clock (see fig. 8).

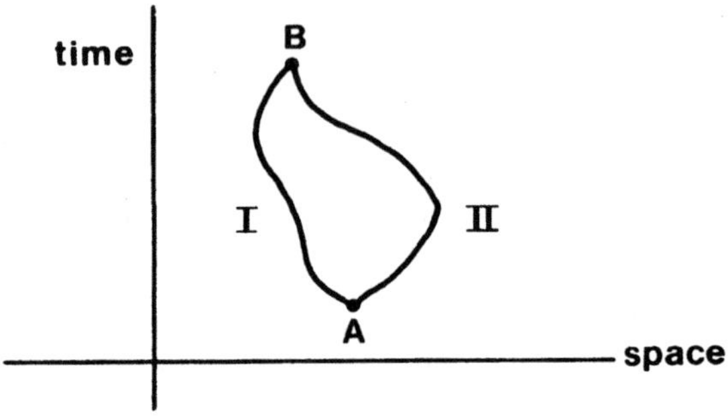

fig. 8

In other words, synchronised clocks at event A are still synchronous at event B, independently of the way they move in between. Parametrising the curves between A and B with absolute time t, as measured by the ideal clocks, makes it possible to state the foregoing in a different way; the integral of dt (the differential of absolute time) between A and B is path-independent: for all paths I and II,

$$\mathrm{I}\int_A^B dt = \mathrm{II}\int_A^B dt \Rightarrow \oint dt = 0 .$$

Absolute time t, therefore, is integrable.

The situation becomes more complicated in the theory of relativity. When starting from relativistic spacetime as the fundamental spatio-temporal entity, there are no longer two separate metrics, one for space and one for time; now, there is but one metric, of the indefinite Riemannian type, for spacetime intervals. In classical physics both the spatial and the temporal distance between two events have objective physical significance, but in RT only the spacetime distance between two events is invariant. The way this spacetime distance is split into a spatial and a temporal distance depends upon the state of motion of the observer.

This results in a quite different behaviour of ideal clocks. Of course, these can no longer be defined as measuring correctly the flow of absolute time, but instead they have to be defined as being free from any disturbances which could slow down or accelerate these clocks (24). Consider again the situation of fig. 8.

According to RT, two ideal clocks synchronised at event A are generally not synchronous when brought together at event B, after having traveled along different routes I and II. Although two observers traveling along I and II make use of identical clocks, they observe that their clocks measure different amounts of time between the events A and B, contrary to what was the case for absolute time. So, a universally valid, absolute measure of time does not exists. This can also be expressed in the following way: parametrise the curves I and II with proper time τ, i.e., the amount of time elapsed along curves I and II as measured by an ideal clock moving along them; then generally, the following inequality holds for I $\neq$II:

$$\mathrm{I}\int_A^B d\tau \neq \mathrm{II}\int_A^B d\tau \Rightarrow \oint d\tau \neq 0 .$$

Thus, proper time τ is not integrable over spacetime trajectories.

Taking the integral of $d\tau$ over all possible timelike trajectories between A and B, there is exactly one trajectory for which the integral takes on a maximum value (25). This trajectory is called the 'geodesic' between A and B. In flat Minkowski spacetime, this geodesic is the straigth line joining A and B, which is the trajectory of an inertial observer for whom both events occur at the same place; in curved spacetime, this geodesic is the trajectory of a freely falling body going through A and B. In the following, I shall restrict myself to flat Minkowski spacetime of STR.

Thus the clock moving along the geodesic between A and B measures more time than any other clock moving along any other timelike trajectory between those events. This is the well known time-dilatation effect of RT. This time-dilatation effect which involves a real difference in time of two clocks should be carefully distinguished from the reciprocal time-dilatation effect which occurs when the time measurements of two observers O and O', in uniform motion relatively to each other, are compared. From O's point of view the clock of O', which he regards to be in motion, is slowed down relatively to his own clock. But because of the relativity of uniform motion, the argument can be reversed, and therefore observer O' will conclude that the clock of O is slowed down, because he regards O to be in motion.

This time-dilatation effect is reciprocal. Because of this reciprocity, the slowing down of clocks in uniform motion is an apparent phenomenon.

I.4.2 A geometrical interpretation of the twin paradox.

The reciprocal character of the time-dilatation effect of clocks in uniform motion is one of the two basic features of the famous twin paradox; the other is the non-integrability of proper time (26). This paradox can briefly be described as follows. Consider the diagrams of fig. 9a and 9b in which the trajectories of two twins T1 and T2 are drawn, as seen respectively by twin T1 and twin T2. In fig. 9a T1 is at rest, while in fig. 9b. T2 is at rest. Suppose that at O both twins synchronise their ideal clocks; what is then the situation at event E? From T1's point of view, fig. 9a, the clock of T2 will be lagging behind at E because the clock of T1, traveling along the geodesic between O and E, will have measured the maximum amount of time between these two events. Therefore, twin T1 will conclude that twin T2 is 'younger' at E.

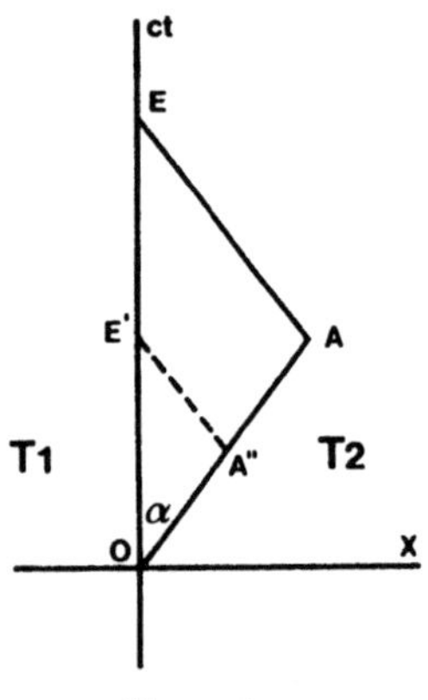

fig. 9a

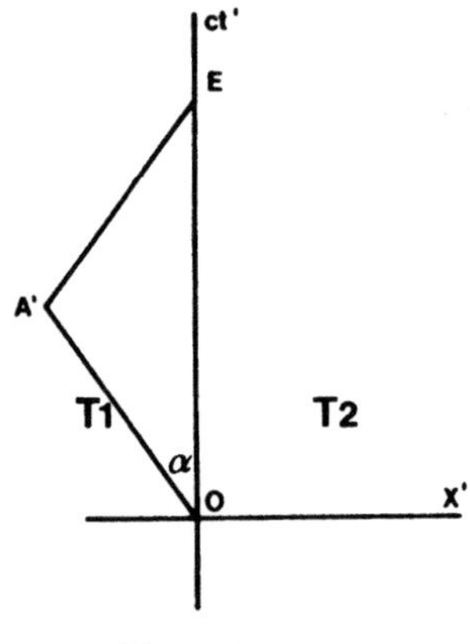

fig. 9b

Although, by itself, this is a rather remarkable conclusion, there is nothing paradoxical about it; it is just a fact about the behaviour of clocks. The paradox arises when the whole situation is described from T2's point of view, fig. 9b, for T2 will reach exactly the opposite conclusion, since he considers himself at rest, and therefore traveling along the geodesic between O and E. Thus he will conclude that twin T1 is 'younger' at E.

RT seems to lead to a contradiction, for precisely, in view of the relativity of motion, it should not matter whether we regard twin T1 or T2 to be at rest. Yet, so the standard argument continues, there is no real contradiction, because the situation between the two twins is not symmetrical at all. Both twins are not physically equivalent, since just one of them 'feels' some accelerations during his trip from O to E and therefore it is possible to determine unambiguously which one of them travels along the geodesic between O and E, and which one along the broken line. If twin T2 feels the accelerations at event A, then the situation of fig. 9a obtains, and T2 will be younger than T1; if T1 feels the accelerations at event A', then the situation of fig. 9b obtains, and T1 will be younger at event E. Thus the twin who has undergone accelerations during his trip from O to E, will be the one who is younger at event E.

This is more or less the argument behind the twin-paradox. Instead of entering into a discussion of the vast amount of literature on this subject, I shall comment on the following three points which might contribute to a clarification of the problems involved:

- 1. the consistency of STR is not endangered by the twin paradox,
- 2. accelerations are not relevant for clarifying the paradox, and
- 3. it is highly misleading to speak of a 'slowing down' of clocks.

ad 1) It is easy to demonstrate that the problem of the twin paradox does not have any consequences for the consistency of STR as a physical theory. According to STR, all inertial observers are equivalent; more particularly, two observers are equivalent if and only if there exists a Lorentz transformation between the two reference frames tied to these observers. Only for these observers, the reciprocal time dilatation effect will occur. But clearly, the transition from reference frame (x,ct), the rest frame of twin T1, to reference frame (x',ct'), the rest frame of twin T2, in fig. 9 does not correspond to a Lorentz transformation, for the straight line OE representing the trajectory of T1 is transformed into the broken line OA'E, whereas Lorentz transformations in Minkowski spacetime transform straight lines into straight lines. Note that the transformation from (x,ct) to (x',ct') is neither the composition of two Lorentz transformations - that would again be a Lorentz transfromation - but is the result of the succesive application of two Lorentz transformations in different spacetime points.

From the foregoing it follows that, if we assume that the two diagrams of fig. 9 represent two descriptions of the same physical situation, the two corresponding reference frames cannot be equivalent. Therefore, whatever time dilatation effect occurs, it will not be reciprocal. In other words, there is no Lorentz symmetry between the two twins, and thus if STR predicts that, e.g., T1 will be younger than T2 at E, then STR does not also predict that T2 will be younger than T1 at E. Any argument leading to that conclusion makes an illicit appeal to the Lorentz symmetry of the twins.

Nevertheless, a paradox remains, for how is it possible to explain that the observed, factual time difference at E in the case of the situation of fig. 9a is equal to the sum of the reciprocal time dilatation effects occurring along OA and AE? If the time dilatation effect for observers in uniform motion is indeed reciprocal, then one might just as well expect that the observed time difference at E is precisely the opposite. I shall return to this problem shortly.

An altogether different problem is to point out the physical cause of the difference in age of the two twins. Very often, this problem is solved by an appeal to the accelerations felt by one of the twins. This brings us to our second point.

ad 2) Apart from the objections one could raise against explicitly introducing an antropomorphism, accelerations cannot physically explain the difference in age of the two twins in E for the following reasons.

In the first place, suppose that the turning acceleration of T2 were at A'' (fig. 9a) rather than at A. If indeed the accelerations at A were the cause of the difference in time at E, then the same time difference should occur at E', since at A and A'' the same accelerations occur. But STR predicts that the time difference at E is greater than the time difference at E'.

Secondly, the whole paradox can be rephrased in such way that it is not necessary at all to refer to physical accelerations. To avoid accelerations is quite simple [Schlegel 1977]. If, instead of one clock moving along OAE (see fig. 9a), two clocks are used, one of which moves uniformly along OA and the second along AE, as illustrated in fig. 10, then accelerations do not occur. In A, clock c3 moving along AE is synchronised instantaneously with clock c2 moving along OA; therefore there is no reason to doubt the equivalence of this three clock version of the twin paradox with the original two clock version. When the readings of clocks c1 and c3 in E are compared, then again the same difference in time will be observed. Yet, there has not been any acceleration of a clock at all.

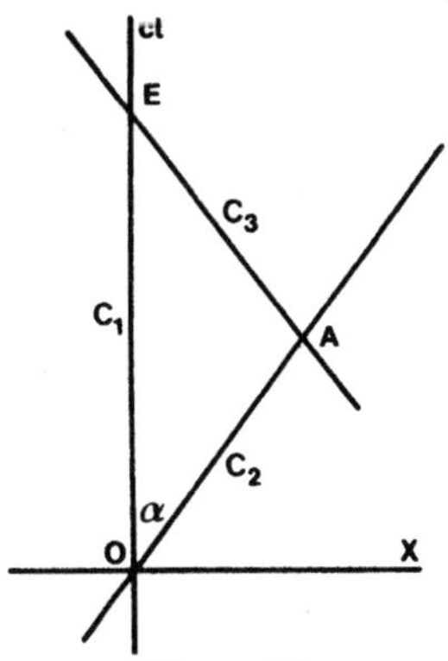

fig. 10

This 'three clock' reformulation of the twin paradox shows therefore that the acceleration of the clock at A in fig. 9a cannot be the physical cause for the different readings of the clocks in E, since the same time difference occurs in the situation of fig. 10 without any accelerations taking place. Of course, if we return from the three clock situation to the one of the twin paradox, with only two clocks, then accelerations must enter the discussion; otherwhise the two clocks would never meet again after having been synchronised at O. But although accelerations necessarily take place, and might possibly enable us to decide which one of the twins is younger at E, they cannot account for the difference in age of the two twins at E.

Finally, if the twin paradox is formulated, not in topologically Euclidean flat spacetime, E^3xE, but in flat spacetime with the topology S^3xE, then it is possible that both twins move along geodesics between O and E; in other words, they will not undergo any physical accelerations [Weingard 1979]. Nevertheless, their clocks will have different readings at E. This again indicates that accelerations are not essential for an explanation of the difference in age of the two twins at E.

Since accelerations cannot explain the time difference at E, there is no need to translate the problem of the twin paradox in terms of GTR, as is, for instance, done by Mittelsteadt [1976, Ch. V], on the basis of the argument that accelerations are involved and that therefore STR cannot adequately deal with this matter (27).

ad 3) In spite of its formal resolvability, the interpretation of the twin paradox remains problematic for the reason indicated at the end of point 1, and moreover, because an effect of uniform motion on the rate of clocks without any intervening physical interaction is counterintuitive [Schlegel 1977, p. 311]. However, as I intend to show now, both enigmas are the result of interpreting the time difference at E in terms of a 'slowing down' of clocks. There are two reasons why it is misleading to speak of a slowing down of clocks, especially within the context of the twin paradox. In the first place, speaking of a slowing down of clocks makes an implicit appeal to the idea of absolute time; secondly, it seduces us into searching for a physical explanation of this slowing down of clocks, in terms of, e.g., physical interaction processes.

The 'slowing down interpretation' of the situation in fig. 9a runs somewhat as follows. The 'real' time difference between the events O and E is the difference measured by twin T1; twin T2 measures a smaller amount of time between the same events, not because the path in time between O and E is shorther along his trajectory, but because his clock is slowed down during the process of measurement. Twin T2 registers a smaller number of ticks between O and E, not because there is less time along his trajectory, but because his unit tick becomes longer during the process of measuring. The concept of an absolute time is thus implicitly retained by virtue of introducing a slowing down process of clocks.

Given this interpretation, the next question is of course: how can we physically explain this slowing down of clocks? What is its cause? Usually, the accelerations at A (fig. 9a) are considered to be the physical cause for this slowing down of clocks. But we have concluded above that accelerations cannot account for the time difference at E; and therefore, they cannot account for the slowing down of clocks. Thus it appears that either uniform motion as such influences the rates of clocks, or it has to be assumed that there exists some hidden interaction process. Both alternatives are equally unattractive, for both must inevitably run aground on account of the relativity of uniform motion, more particularly with regard to the situation in fig. 10, on account of the reciprocity of the slowing down of the clocks c1 and c2, c2 and c3 and finally c3 and c1.

Without the notion of absolute time, the slowing down interpretation looses its meaning, but abandoning absolute time also implies that there is nothing counterintuitive about two clocks measuring different amounts of time between the same events without any intervening physical interaction with these clocks. In this respect, the fact that the time difference at E is independent from the constitution of the clocks used for measuring the time between O and E, is of paramount importance. This fact is hard to explain within any interaction interpretation of STR, for in that case one would expect that these interactions act like 'differential forces' [Reichenbach 1957] for clocks of different constitution. This independence strongly suggests that we are dealing here with matters of a geometrical nature [Weingard 1981]. An analogy with space clarifies this geometrical interpretation of the twin paradox.

Consider again the situation of fig. 9a, but now interpreted as paths in space, not in spacetime. The triangle inequality entails that the distance between O and E measured along OE is shorter than the distance between the same points measured along OAE (28). This is a fact about geometry; there is no need to explain it by arguing that certain physical interaction processes cause the meter stick along OAE to shrink and that thus, a greater distance will be measured along that path. Clearly, this would be absurd. But precisely this line of argument is followed in the slowing down interpretation; it looks for an explanation of the fact that synchronised clocks will not stay synchronous in the future. This, however, is a basic fact about the behaviour of standard clocks in the same way that the triangle inequality is a basic fact about the behaviour of meter sticks. Therefore, any attempt to make the time difference at E plausible, for instance in terms of signal procedures as proposed by Bohm [1965, Ch. XXX], is in vain: they show only the consequences of this basic fact about the behaviour of standard clocks, which is codified in the geometrical structure of the spacetime manifold and belongs to the axiomatic structure of STR [Weingard 1981, p. 224].

This analogy with space also reveals why the sum of the reciprocal time dilatation effects along OA and AE with respect to OE, in spite of its reciprocity, is exactly equal to the real time difference measured at E. Consider the projections of two meter sticks, lying along OE and OA, upon each other; there appears to be a reciprocal shortening of their length by a factor $\cos\alpha$. Elementary geometry shows that the real difference in length between OE and OAE is determined by the reciprocal shortening factors of the meter sticks. Similarly, the reciprocal time dilatation effects are related to the time differences that are really measured. There is nothing paradoxical about it.

I.4.3 The Unity of Time.

The rejection of the notion of absolute time has brought about yet other kinds of problems, such as the problem of the epistemological status of the congruence of time intervals. Within the Newtonian framework, the congruence of time intervals is an intrinsic property of those intervals. The equality of time intervals is determined by the flow of absolute time and, as the latter is independent of anything external, the former is too; whether or not two time intervals are equal in length is ultimately a question of facts. The rejection of absolute time, however, changes the situation. Reichenbach has convincingly shown that [1957, p. 116]:

> the equality of successive time intervals is not a matter of <u>knowledge</u> but a matter of <u>definition.</u> As for spatial congruence, a cer-

tain rule must be laid down before the comparison of magnitudes is defined.

Speaking of the equality of time intervals is meaningful only after the appropriate 'coordinative definitions' have been given. Of course, our choice of a particular coordinative definition is conditioned by our desire to make the description of physical reality as simple as possible. But, as Reichenbach [op. cit. p. 117] remarks, this "simplicity has nothing to do with truth, since it is merely descriptive simplicity." In other words, the statement that two time intervals are equal in length can only be 'conventionally true' [Sklar 1977, p. 123].

This analogy in the epistemological problems concerning the metric of space on the one hand and the metric of time on the other, is not surprising since in the spacetime approach, both coalesce into the spacetime metric. In both instances, the same kinds of problems are encountered.

With respect to time, however, Mittelstaedt [1976, Ch. I and II] has pointed out a particularly interesting problem which he refers to as the problem of the Einheit der Zeit, the 'Unity of Time'. This problem is also related to the abandonement of the notion of absolute time. To Newton it was self-evident that there exists just one fundamental metric for time: the intrinsic metric of absolute time. Physical processes of whatever kind could provide a more or less accurate 'sensible measure' thereof. A 'true' sensible measure of absolute time could only be reached in the case of a perfectly isolated, completely undisturbed periodical system which would constitute an ideal clock. In a certain sense, all physical processes had to obey, according to Newton, the rhythm of absolute time: an ideal clock, of whatever nature (mechanical, gravitational etc.) could provide an exact measure of the unique, fundamental metric of absolute time.

As soon as the idea of an absolute time is abandoned, the situation becomes more problematic. There is no a priori reason to assume that different kinds of physical processes will lead to one metric for time. Consider, for instance, the time scales defined by, respectively, a mechanical, an electromagnetical and a gravitational periodical system. Empirically, we know that these time scales 'match' to a very high degree of accuracy. In other words, a mechanical, electromagnetical and gravitational clock, once they are synchronised, stay synchronous to a very high degree of accuracy. Newton would interpret this in terms of one fundamental metric for time. However, there is no a priori reason to assume that these different time metrics do in fact harmonise [Reichenbach 1957, p. 117]. Usually, the assumption that indeed, once synchronised, these different clocks remain synchronous, is taken for granted. The hypothesis that all kinds of independent physical systems - mechanical, electromagnetical, gravitational, thermodynamical, quantum mechanical etc. - ultimately lead to one fundamental metric for time is called by Mittelstaedt the principle of the 'Unity of Time' (die 'Einheit der Zeit'). This principle of the Unity of Time not only applies to the metrical structure of time, but also to the order and topological structures (29).

The principle of the Unity of Time has mainly a heuristic, methodological function. This can be seen in the following way [Mittelstaedt 1976, p. 38-39]. Suppose that after carefully comparing the metrics ascribed to time on the basis of inertial and gravitational systems, the two resulting metrics of time, T^i and T^g, appear to be non linearly related. (Linearly related time functions are not of interest since they lead to the same congruence classes on the set of time intervals.) This means that the most 'natural' unit of time for gravitational processes varies relatively to the most 'natural' time unit for inertial processes (or vice versa), since these units of time have been chosen so that the equations describing the behaviour of gravitational and inertial systems look as simple and intelligible as possible. Usually, T^i and T^g are chosen so that the inertial rest mass m, respectively the gravitational constant G do not vary in time. But if T^i and T^g are non linearly related, then it is impossible to choose the metric of time in such way that both m and G are constant in time. Relatively to T^i, G will be time dependent and relatively to T^g, m will vary in time. In such situations, the principle of the Unity of Time functions as a methodological rule stating that we should try to find out which metric of time is the more fundamental one; in other words, we should attempt to discover which type of phenomena, gravitational or inertial phenomena, is to be regarded as more basic. Once this issue is settled, the next step is of course to explain why, if e.g. the inertial phenomena are considered to be more fundamental and therefore, T^1 is taken to be the more basic metric of time, the gravitational constant G changes relatively to the inertial time T^1.

Undoubtedly, most physicists would not object to the use of the principle of the Unity of Time as a methodological rule. Yet, this methodological rule asks for a legitimation of its use. According to Mittelstaedt [op. cit. p. 39], the principle of the Unity of Time is based upon the more general principle of the 'Unity of Physics' (die 'Einheit der Physik') which refers to the idea that all known theories of physics can be derived from, or encompassed in a single, fundamental theory. This fundamental theory, if it exists, determines the ultimate metric of time. So, it is an extremely reductionistic position that grounds the belief in the validity of the principle of the Unity of Time. Mittelstaedt [op. cit. p. 39-40] remarks, however, and I fully agree with him, that the idea of a fundamental metric of time could be an illusion, because the Unity of Physics, upon which it is based, could be a fiction.

I know of just one theory in which the principle of the Unity of Time is rejected as a methodological rule. This is the case for Milne's kinematic relativity theory [Milne 1948] in which he uses two different time scales t and τ which are logarithmically related to each other. In the τ-scale, Newtonian physics is valid and so it is "finally to be identified with the 'uniform' time of physics" [op. cit. p. 36]. The t-scale, on the other hand, is to be applied in cosmological matters: "nature, as it were, knows only t-measures in these cosmological questions" [op. cit. p. 230]. In the t-scale, the gravitational constant G is time dependent. According to Milne, these two time scales are equivalent and constitute two different languages; physical reality can be described in both languages. The description of phenomena at the cosmological level leads to congruence classes of time intervals which are different from the congruence classes of the time function used in ordinary physics. In Milne's theory, 'local' and 'global' time have different metrics, but both are equivalent, and thus, neither one is more fundamental for the description of physical reality. Here the principle of the Unity of Time as a methodological rule is not obeyed; no attempt is made to discover which metric is more basic than the other. Both metrics are accepted as equally fundamental.

I.5 Conclusion

The transition from the concept of absolute time, as introduced by Newton in his Principia, to the notion of time as used in the spacetime approach of RT, represents a drastic step in the physicist's conception of physical time. RT prompted a profound revision, at the order, respectively, topological and metrical level, of long-standing and seemingly self-evident ideas about time, anchored unerkannt im Unbewussten ('unrecognisedly in the unconscious') as Einstein [in Schilpp 1969, p. 52] expressed it. We do not necessarily have to think of physical time as a continuous straight line. The structure of time is not fixed a priori, as Kant thought it was (30). We have come to realise that, besides the standard picture of time, there exist other possibilities for the structure of physical time.

I have reviewed some of the epistemological problems engendered by this insight. In view of the failure of the Kantian a priori conception of the structure of time, the most obvious alternative is to look for 'a posteriori evidence' in nature in order to determine the actual structure of physical time. We have seen, however, that the matter is not altogether as simple as that. The known empirical data fit different conceptions of time; but even worse, it appears that whatever the future empirical data may be, these data will be compatible with different time concepts. This leads to a Poincaréan conventionalistic attitude [Poincaré 1952] that regards different alternative time concepts as roughly equably viable, and maintains that our choice of one of them is essentially free. Strange as this may sound, this closes the circle and brings us back to a kind of a priori conception of the structure of physical time: a complete determination of the structure of physical time is not possible upon the basis of what is empirically given. But this conception is not a priori in the Kantian sense, since the properties of time are considered to be of a contingent nature (31).

Looking at the great impulse given to the study of time by RT, it is rather remarkable that quantum mechanics, that other paramount achievement of twentieth century physics, had little or no influence upon the physicist's conception of time. Quantum mechanics is not based upon, nor does it contain a new theory about the structure of physical time. Time in quantum mechanics is usually considered to be homeomorphic to the Euclidean line: the idea of time as a straight line. However, time plays a rather obscure role in quantum mechanics. But here we broach upon the subject matter of the next chapter which concerns the 'geometrical role' of time in physical theories.

NOTES

1 In Ch. II, the 'space and time' and the 'spacetime' description of physical reality will be compared. There too the conditions under which a translation between both is possible, will be discussed.

2 For a more detailed exposition, see Sklar [1977, p. 46 ff].

3 See the discussion in sect II.8; cfr. also Earman [1970].

4 I assume here and in the following the so-called Einstein convention for the synchronisation of distant clocks; cfr. Grünbaum [1973, Ch. 12].

5 In contrast to this cyclic order, the order of the points on the Euclidean line is often called 'linear' order.

6 See, for instance, Grünbaum [1973, Ch. 7], Earman [1972b], Winnie [1977] and Denbigh [1981, p. 30 ff]. For a history of the causal theories of time, see Mehlberg [1980].

7 The derivation of the metric structure of Minkowski spacetime from the causal structure of STR is rather remarkable for the following reason. The conformal structure of a spacetime reflects completely the causal structure of that spacetime, but generally different metrics are compatible with a given conformal structure. Therefore the determination of a unique metric on the sole basis of the conformal structure (i.e. causal structure) of the spacetime manifold is in general not possible. According to Winnie, this is not the case for Minkowski spacetime [Winnie, 1977, p. 195-196].

8 More precisely: a subset A of M is called 'open' if for every point x in A there is an ε-ball $B=\{y|$distance between x and y is smaller than $\varepsilon\}$, $\varepsilon>0$, which contains x and is at the same time wholly contained in A.

9 Cfr. Winnie [1977, p. 157]. Strong causality roughly means that every event has arbitrarily small neighbourhoods to which no timelike curve returns once it has left that neighbourhood.

10 This is not always done; for instance, Reichenbach [1957, p. 136] writes that "the first topological determination of time deals with <u>time order at the same point</u>; Lucas' [1973] chapter on "The topology of

time" mainly deals with temporal order, and according to Denbigh [1981, p. 21] there is a relation between the fact that there is only a single temporal order relation and the one-dimensionality of time.

11 A 'time slice' is a spacelike hypersurface of M which is a closed subset of M without boundaries.

12 For the notion of covering spacetime, see Sklar [1977, p. 312 ff]. For a discussion of observationally indistinguishable spacetimes see C. Glymour, "Indistinguishable space-times and the fundamental group" in Earman [1977a].

13 Roughly, spacetime A with closed time is the result of 1) 'rolling up' the covering Minkowski spacetime B with linear time, and 2) identifying the points O, O', O'',..., and a', a'', a'',...etc. In the transition from a spacetime M to its covering spacetime M', one spacetime point of M, e.g., a in fig. 6, is replaced by an infinite number of spacetime points in M'. Here the difficult problem of the identification of spacetime points arises.

14 The reason is that the Gödel spacetime is simply connected and therefore its own universal covering. For more details, see Earman [1977b], Glymour and Malamant in Earman [1977a], and Sklar [1977].

15 A metric space is called 'Cauchy complete' iff every Cauchy sequence in M converges to a limit point in M. A sequence s1, s2, s3 etc. of points in M is called a 'Cauchy sequence' iff for each positive ε there exists an integer N such that the distance between si and sj is smaller than ε whenever both i and j are greater than N [McCarty 1967, p. 163].

16 Cfr. Hurewicz [1941, p. 4]. Following this definition, the number of dimensions is always an integer. Other definitions can be introduced which allow for non-integer values for the number of dimensions (the so-called 'fractal' dimensions); cfr. Mandelbrot [1977, p. 296-297], who briefly discusses the possible use of 'fractal times' in physics. I will not consider these fractal times here.

17 See, for instance, Melyukhin, "A philosophical evaluation of modern ideas concerning the properties of space and time in the microworld", in Kuznetsov [1965]; also Lucas [1973, p. 32] and the references given there.

18 Sometimes it is argued that in case the total energy in the universe is finite, Heisenberg's uncertainty principle $\Delta E \Delta t \geq \frac{1}{2}\hbar$ implies that there is a minimum in the accuracy of time measurements.

19 For discrete time there is no reason to assume that the length of an arbitrary time interval is equal to a multiple of a fundamental time unit. For instance, dense time is discrete, but the notion of next instant and a fortiori of the smallest time distance between instants lying next to each other, is meaningless.

20 According to Sklar [1977, p. 301], temporal orientability is a notion which is meaningful only in the spacetime approach; this, however, is not true.

21 For the notion of time reversal invariance of physical laws, see Ch. III.

22 The CPT* theorem states that any process must be invariant under a combined charge (i.e. particle-antiparticle) conjugation C, space-inversion P and time reversal T; cfr. Sakurai [1973, p. 173-174]. See also Ch. III.

23 For the definition of a distance metric, a measure metric and a Riemannian metric, see Grünbaum [1973, p. 468 ff].

24 In Reichenbach's terminology [1957, p. 13], they have to be free of 'differential forces'.

25 In curved spacetimes, the integral may take on a maximum value for more than just one trajectory.

26 The discussion of the twin paradox in this section is a slightly modified version of Kroes [1983].

27 Moreover, it is a 'myth' that STR cannot deal at all with accelerations; cfr. Bondi [1967, p. 53].

28 Note that in space the distance between O and E along OE is shorter than along OAE, whereas the spacetime distance along OE is greater than the spacetime distance along OAE; this is due to the indefinite metric of spacetime.

29 Mittelstaedt's interpretation of the Unity of Time at the topological level in terms of a unique direction of time for all phenomena, is rather misleading, for the direction of time is not a topological property (see Ch. III). It is more coherent to interpret the unity of topological time as the assumption that different kinds of physical processes will ultimately lead to the same topological structure for time.

30 Kant [1956, p. 74 ff] and Kant [1969, p. 32 ff]. For Kant the a priori character of the structure of time implied that this structure was universally and necessarily valid.

31 For a discussion of the difference between the 'contingent a priori' and the 'Kantian a priori', see Stegmüller [1975, p. 221 ff].

CHAPTER II

THE GEOMETRICAL NATURE OF PHYSICAL TIME: PARAMETER TIME AND COORDINATE TIME

II.1 Introduction

The topic of this chapter no longer concerns the 'internal' structure of time, but is related to the role which time plays in physical theories and to questions about the relations between time and other fundamental concepts appearing in those theories.

In order to elucidate the meaning of the expression 'the geometrical nature of time', let us focus our attention on the description of motion and ask how time functions in the different ways in which motion is described in physics. Almost any motion -'motion' taken in the broad sense of 'change', e.g. the motion of a particle in physical space or the evolution of a thermodynamical system - is usually represented in physics by way of a 'trajectory' in some kind of (abstract) space (see fig. 11) (1). Ordinarily, these trajectories are given by 'parametrised curves'.

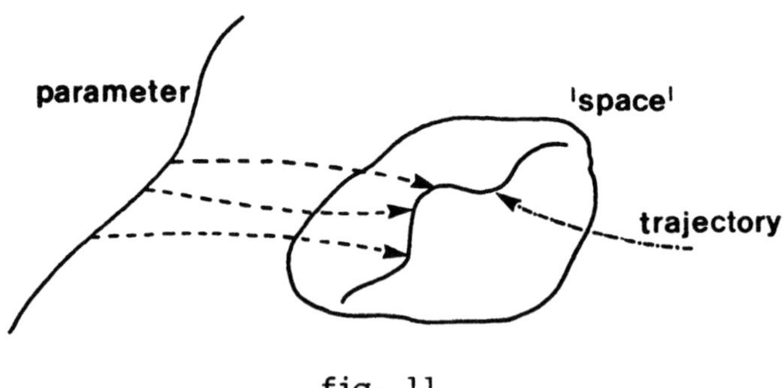

fig. 11

In general, not every curve in the space in which the motion is represented can be interpreted as corresponding to a physically realisable motion; therefore, restrictions have to be imposed upon the types of curves that can be called 'trajectories'. Some of these restrictions are of a very general nature, such as those concerning the continuity and differentiability of the curves; others depend, as we shall see later on, upon the nature of the (abstract) space and upon the particular physical theory involved.

The (abstract) spaces used for the description of the evolution of physical systems, are mostly taken to be differentiable manifolds (2). Therefore, the standard scheme for the description of motion in physics is constituted by a parametrised curve in some differentiable manifold (see scheme 1). Because we presuppose continuity and differentiability with regard to the spaces and the curves, the use of this standard scheme puts strong limitations upon the type of physics we are dealing with.

$$\begin{array}{cccc} C: & I & \longrightarrow & M \\ & \omega & & \omega \\ & t & \longrightarrow & c(t) = m(t) \end{array}$$

scheme 1

The nature and the structure of the manifold M is closely connected with the nature of the physical system under consideration. For example, usually the motion of a material particle is represented by a trajectory in space (E^3) or spacetime ($E^3 x E$), both of which have a specific metric, whereas the phase spaces in which the evolution of thermodynamical systems is represented are taken to be affine spaces. The same applies to the parameters: while in the first case, time is usually chosen as the parameter, in the second temperature may sometimes play that role. Note that there is no unique way for describing the evolution of a particular physical system; neither the parameter nor the space is fixed a priori in a definite way. Different choices for the parameter and the space are in general possible for the representation of the evolution of the same physical system. One of the aims of this chapter is precisely to compare and evaluate two different ways of describing the motion of a material body (conceived as a point particle).

From now on, I shall confine myself to the description of change of place of a (point) particle, that is to 'locomotion'. Time as well as space, usually play a fundamental role in the description of this kind of motion: physical space or spacetime is taken to be that in which the motion takes place, while time is used as a parameter, because at each moment of time, the position of the particle in space or spacetime is uniquely determined. But as Dingle [in Schilpp 1969, p. 540] has rightly pointed out, it is by no means necessary to measure and represent change of place "by the space covered by the moving body in a given time". "Having no preconceptions in the matter at all", we could have chosen, according to Dingle [ibidem], to construct some form of speedometer or to use "the displacement of spectrum lines known as the Doppler effect". In taking time and space as the fundamental concepts for describing motion, physicists follow Galileo who was one of the first to recognise the importance of time within this context. Thus, beside the standard interpretation of the manifold M as corresponding to physical space or spacetime, and of the parameter t as corresponding to time, other interpretations are in principle possible.

Within the context of this standard scheme for the description of motion, the expression 'the geometrical nature of time' refers to the nature of the relationship between time and the basic geometric entities involved in this scheme: the manifold in which the motion is represented and the parameter. Time can be related in two fundamentally different ways to the manifold in which motion is represented. If it is part, a submanifold, of that manifold, then it functions as a coordinate; if it is 'external' to the manifold, it functions as a parameter. This leads to the distinction between parameter time and coordinate time each of which plays a different geometrical role in the description of motion [Scheurer 1979].

This chapter deals mainly with the geometrical role of time in the kinematics of classical and relativistic mechanics; also the geometrical role of time in quantum mechanics will briefly be considered.

The transition from classical Newtonian to relativistic Einsteinian mechanics is marked by a transformation in the geometrical nature of time, at least in the standard formulations of both theories. In Newton's description of motion, time functions only as a parameter. Motion takes place in absolute space and is parametrised by absolute time; to each moment of absolute time, there corresponds a unique position of the particle in absolute space. Moreover, absolute space and time are considered to be completely independent. In the description of motion this independence is reflected in the fact that the parameter and the space manifold are completely independent of each other.

In the standard spacetime formulation of relativity theory, this picture changes thoroughly. Now time, besides being a parameter in the form of proper time, is also treated as a coordinate. This change in the geometrical nature of time, from parameter time in classical mechanics to the couple of parameter and coordinate time in relativistic mechanics, is parallelled by a transformation of the notions of space and time into the single concept 'spacetime'. This again has had a thorough influence upon the physicist's conception of nature; not (material) objects, but events are regarded to be the ultimate constituents of the external world, and these events simply are, they cannot be said to be happening. This position is most clearly stated by Weyl [1963, p. 116]:

> The objective world simply <u>is</u>, it does not <u>happen</u>. Only to the gaze of <u>my</u> consciousness, crawling upward along the life line of my body, does a certain section of this world come to life as a fleeting image in space which continuously changes in time.

It goes without saying that such a far-reaching change in the physicist's conception of nature gave rise to many philosophical problems especially with regard to the notion of physical time.

In this chapter I shall undertake a critical examination of some of these philosophical problems while taking into account:

i) the different geometrical role that time plays in the space and time and the spacetime description of motion, and

ii) the fact that Newtonian and Einsteinian mechanics can be formulated in non-standard ways: relativity theory, at least the special theory, can be formulated within the space and time approach with the sole use of parameter time, and Newtonian mechanics can be formulated within the spacetime approach with parameter and coordinate time.

It will be shown that the widespread idea that relativity theory necessarily leads to a static conception of the physical world is based upon a misunderstanding of the geometrical role of time in that theory, and that certain philosophical problems associated with the theory of relativity are not related to that theory itself, but to the spacetime formalism in which it is usually formulated. Thus, one ought to be very careful to distinguish between the philosophical problems belonging to the theory of relativity proper, and those connected with the use of the spacetime formalism. It will also be argued that the distinction between parameter time and coordinate time deserves more attention in the study of time than it has received sofar; in particular, I shall oppose Reichenbach's claim that the treatment of time as a coordinate in relativity theory is only relevant for mathematical physics but not for the study of time.

I shall begin with a brief discussion of the notions of parameter and coordinate.

II.2 Parameter and coordinate

Let M be an n-dimensional differentiable manifold and I the standard open interval on the reals R. A parametrised curve c(t) on M, with parameter t, is a mapping from I into M:

$$\begin{array}{lccl} C: & R \supset I & \longrightarrow & M \\ & \rotatebox{90}{$\in$} & & \rotatebox{90}{$\in$} \\ & t & \longrightarrow & c(t) = m(t) \end{array}$$

A parameter therefore is simply a variable ranging over the standard open interval of the reals. In the following we will take c(t) to be a differentiable curve, which means that the mapping C is differentiable.

Given a curve C in M, it can be parametrised in an indefinite number of different ways. In general, all these parametrisations are equivalent; the choice of a particular one is arbitrary. From a physical point of view, however, the class of allowable reparametrisations of a curve is very important. For instance, within

Newtonian mechanics, only linear transformations of parameter time (t' = at + b, a and b constants) are allowed.

It is not necessary to recall here the precise definition of a coordinate system on M. For our purposes, the following rough characterisation is sufficient. Given open subsets U of M and W of R^n, a coordinate system x on U consists of a set of functions $(x_1, \ldots\ldots\ldots, x_n)$ from U to W such that x is a homeomorphism (a one-to-one bicontinuous map) between U and W. These functions assign to each point m of U its coordinate values $x_1, \ldots\ldots\ldots, x_n$. A coordinate function, therefore, is given by a mapping from M to R; it is a numerical function on M:

$$x_i : U \subset M \longrightarrow V \subset R \qquad x : U \subset M \longrightarrow W \subset R^n$$

$$m \longrightarrow x_i(m) \qquad m \longrightarrow (x_1(m), \ldots, x_n(m))$$

Since a coordinate system is a homeomorphism, all the topological properties of the domain U and the codomain W are the same. Locally, there is just one topological property that completely determines the structure of the manifold: the number of dimensions [Thirring 1977, p. 12] (3).

The foregoing definitions clearly show the difference between a parametrised curve and a coordinate function. A parametrised curve is given by a mapping from the reals to the manifold, whereas a coordinate is a mapping from the manifold to the reals; they are dual mappings (4):

$$R \xrightarrow{\text{parameter}} M \qquad M \xrightarrow{\text{coordinate}} R$$

The difference between them gets obscured in case M = R, for then, a coordinate and a parameter are given by a function from R to R. Suppose furthermore that the mapping producing the parametrisation of R is a homeomorphism; then any difference between both notions disappears (5). Under these special circumstances, the notions of a parametrised curve and a coordinate function have the same meaning: a coordinatisation of the reals is at the same time a parametrisation of the reals. In general however, the two should be carefully distinguished.

Since a parametrised curve is given by a mapping, each parameter value corresponds to a unique element of M. In terms of the description of motion, this means that we can 'follow' the motion of a particle in M: for each parameter value it is possible to locate the particle unambiguously in M. Only in very exceptional cases, the same is possible with the help of a coordinate function. Given a coordinate system on M and a curve c(t) in M, the latter can be 'parametrised' by one of the coordinate functions x_i, iff the function x_i is a bijection between the points of the curve and the reals. If so, the parameter values for the points on the curve can be chosen so as to be equal to the corresponding x_i coordinate values. In this way, it becomes possible to identify a parameter and a coordinate, a fact that will be important later on for the problem of finding a translation between the spacetime and the space and time description of motion.

Another important aspect of a parameter is that it is 'extrinsic' or 'external' to the manifold; the parametrisation of a curve in M is in principle completely independent of the structure of M. In case M is a metric space, however, a curve in M is very often parametrised by the arclength along the curve from a fixed point. This, for instance, is standard practice in relativity theory where the parametrisation of a curve in spacetime is given by proper time (i.e. the spacetime distance along the curve).

Finally, the following point concerning the physical significance of coordinates deserves our attention. In the general theory of relativity the idea that differences in space and time coordinates have a direct physical significance in terms of measurements with rigid rods or standard clocks, must be put aside [Einstein 1952, p. 117]. As a result, differences in coordinate time values between events have generally no direct physical significance; they are not a direct measure of the time elapsed between those events. The values of parameter time however are chosen so as to have direct physical significance. That they can so be chosen at all, follows from the fact that the parametrisation of a curve in a manifold is in principle independent of the structure of M.

The parametrisation of the trajectory of a particle is fixed physically by the following convention: the difference in parameter values for two points on the trajectory must be equal to the time elapsed between the two corresponding events as measured by a comoving standard clock, i.e. a clock that travels along with the particle. This determines the parametrisation completely except for a multiplicative constant (the unit of time of the standard clocks), and an additive constant (the origin of parameter time). It is rather obvious to choose the additive constants for different trajectories so that all events with parameter value zero are simultaneous. But then events on different trajectories with the same parameter value are in general not simultaneous, since that would imply a strong presupposition about the behaviour of clocks, which cannot be justified a priori and which moreover is contrary to the actual behaviour of clocks (see p 40 ff).

The foregoing implies that, whenever both parameter and coordinate time are used for the description of motion, as in the spacetime formalism, parameter time is the more fundamental notion of time because it has direct physical significance.

I shall now turn to an exposition of the role of parameter time and coordinate time in Newtonian and Einsteinian physics.

II.3 Parameter and coordinate time in Newtonian physics

II.3.1 Newtonian space and time: absolute time as parameter.

In the Principia, Newton [1974, p. 6-7] introduces the notions of absolute and relative time, space and motion:

> Absolute, true, and mathematical time, of itself, and from its own nature, flows equably without relation to anything external, and by another name is called duration: relative, apparent, and common time, is some sensible and external (whether accurate or unequable) measure of duration by the means of motion, which is

> commonly used instead of true time; such as an hour, a day, a month, a year.
> Absolute space, in its own nature, without relation to anything external, remains always similar and immovable. Relative space is some movable dimension or measure of the absolute spaces;...
> Absolute motion is the translation of a body from one absolute place into another; and relative motion, the translation from one relative place to another.

These concepts determine the framework within which Newton constructs his kinematics and dynamics. Newton clarifies the distinction between the notions of absolute and relative time, space and motion in the following way. In common affairs, we can without any inconvenience use the relative measures of time and space. "But", he continues, "in philosophical disquisitions we ought to abstract from our senses, and consider things in themselves, distinct from what are only sensible measures of them" [op. cit. p. 8]. 'Relative' in this Newtonian sense is not to be confused with 'relative' in the specific Einsteinian sense, where it refers to the fact that space and time intervals are always measured relatively to a certain reference frame; nor should 'relative' be confused with 'relational': relational theories about space and time (spacetime) deny that space and time (spacetime) are entities existing independently of matter (events), but regard them as attributes of matter (events). These relational theories concern the ontology of space and time.

Of special interest is what Newton has to say about the relation between absolute space and absolute time. He is very clear on this point; they are completely independent from each other. Each exists in its own right; the phrases "absolute time....without relation to anything external" and "absolute space....without relation to anything external" show this beyond any doubt. Space and time are separate entities; space is 'external' to time and vice versa. Therefore there is no question at all of blending together space and time into a four-dimensional spacetime as in the standard formulations of relativity theory. Consequently, the flow of absolute time is independent from both the place and the change of place (the state of motion) in absolute space: absolute time is universal.

For Newton, the external world consists of a three-dimensional Euclidean absolute space filled with material particles. Those material particles persist throughout the flow of absolute time. To each instant of absolute time, there corresponds a certain distribution of the particles in absolute space, and the change of place in absolute space from one moment to another is called absolute motion. Thus, the concept of absolute motion involves both absolute space and time, but only as independent entities.

Within this context, the scheme for the description of motion of page 59 takes the following form (6):

	Absolute time		Absolute space
C:	E	----->	$M = E^3$
	t	----->	$m(t) = (x(t),y(t),z(t))$

For Newton, absolute space was necessarily Euclidean (7). In this scheme, any continuous differentiable curve in M represents a physically realisable motion. Particles in uniform (inertial) motion are represented by straight lines, and particles at rest by a unique point in M.

It is a remarkable feature of the space and time description of motion that the trajectories of particles at rest, contrary to particles in motion, are given by 'degenerated' curves in M, i.e. by unique points. This feature of the space and time formalism favours a physical interpretation of nature in which motion and rest are regarded as two fundamentally different states of a particle. In the history of physics, this point of view was standard until the 17th century under the influence of Aristotle who regarded motion and rest as opposites [Physica, 226^{b}]. In the modern vision on motion and rest, there is no fundamental difference between them because rest is regarded as a state of motion but with zero velocity (8).

Within Newtonian mechanics, it is not possible to distinguish dynamically between particles at rest and particles in uniform motion; therefore a (restricted) principle of relativity is valid [Einstein 1952, p. 111]. The corresponding class of allowable reparametrisations of curves in space is determined by the well-known (generalised) Galilei transformations which transform inertial motions into inertial motions:

$$t' = at + b$$
$$x' = cx + dt + e,$$

where a,b,c,d and e are constants. The class of 'degenerated' curves in M, corresponding to particles at rest, is not left invariant under these transformations. Therefore, from a dynamical point of view, the difference between motion and rest (reflected kinematically in the shape of the curves) disappears.

The choice of absolute time as the parameter is rather obvious. First, to each moment of absolute time there corresponds a unique place of the particle in absolute space; this is a conditio sine qua non for using absolute time as parameter. Secondly, absolute time satisfies the requirement of being independent from the manifold in which the motion is described. Finally, differences in absolute time are the same for all observers and therefore this choice makes the parameter a physically invariant quantity. In other words, given ideal undisturbed clocks which measure correctly the flow of absolute time, each observer equipped with such a clock will measure the same difference in absolute time between two events A and B, independently of his place or state of motion. Consequently, those ideal clocks behave very nicely: once two such clocks are synchronised, they will stay synchronous forever, regardless of their motion (see pp 39-40).

The foregoing makes it possible to parametrise trajectories in such way that events with the same parameter value are simultaneous. Assuming that the events corresponding to the origins of the parametrisations are simultaneous, the parametrisation of trajectories by absolute time guarantees that the same parameter values are assigned to simultaneous events. The possibility of such a coherent invariant parametrisation of different trajectories is a characteristic feature of Newtonian absolute time.

II.3.2 Newtonian spacetime: absolute time as parameter and coordinate.

Ever since the elegant 'reformulation' of the special theory of relativity in the four-dimensional spacetime formalism by Minkowski, physicists have tried to reformulate Newtonian mechanics and gravitation theory into the same formalism. The important contributions in this field made by Frank [1909], Cartan [1923, 1924] and Friedrichs [1927], were followed up, about two decades ago, by the work of Havas [1964] and Trautman [in Hoffman 1966]. By reformulating Newtonian physics in the spacetime formalism so that its predictions are identical with those that can be derived from its original formulation, a common 'universe of discourse' is created for both classical and relativistic theories. On the one hand, this is very interesting for the physicist because it puts him in a better position to appreciate the exact similarities and differences between both theories. On the other hand, it is also of interest for the philosopher of science; these reformulations of different theories into a common language, may throw a new light upon the issues concerning the relation between successive theories within a particular domain of inquiry (e.g. the problem of reduction of theories or of the incommensurability of theories). These reformulations also deserve our attention because they can contribute to a better understanding of some of the philosophical problems about time connected with the transition from Newtonian to Einsteinian mechanics.

I shall begin with a closer look at the structure of Newtonian spacetime; therafter I shall turn to the description of motion in Newtonian spacetime.

The structure of Newtonian spacetime.

Starting from a spacetime manifold, the problem of recovering the Newtonian analogues of absolute space and time presents itself. This can be achieved rather easily by assuming that spacetime possesses a particular structure. The key ideas behind this structure are summarised by Weyl as follows [1963, p. 95]:

> A definite structure is already ascribed to the four-dimensional extensive medium of the external world if one believes in a severance of the universe in the sense that it is objectively

> significant to say of any two different events, narrowly confined in space-time, that they are happening at the same place (at different times) or at the same time (at different places). All simultaneous world-points form a three-dimensional stratum, all world-points of equal location a one-dimensional fiber. According to this view we may describe the structure of the world as possessing a stratification whose layers are traversed by fibers. (Through each world-point runs one stratum and one fiber; any one fiber intersects a stratum in but a single world-point.)

The beginning of a reconstruction of absolute space and time is made by assuming that 'being simultaneous with' and 'being at the same place' have an objective, physical meaning. But this is not sufficient, for it is also necessary to introduce a metric for space and time [op. cit. p. 96]:

> One attributes furthermore to space and time a metrical structure by assuming that equality of time intervals and congruence of spatial configurations have an objective meaning..

The metric of each stratum is of course Euclidean, as was the case for absolute space. Note that Weyl refers to two different metrics, one for space and one for time, instead of a single metric for spacetime intervals. The absence of a spacetime metric in Newtonian spacetime is one of the most conspicuous differences from relativistic spacetimes.

Like Weyl, Misner et al. also use stratification (9) of the spacetime manifold as a tool for obtaining the analogues of absolute space and time [Misner 1973, p. 291]:

> How do the absolute space of Galileo and Newton, and their absolute time, fit into [....] "Newtonian spacetime"? The key to this fit is stratification; stratification produced by the universal time coordinate t.
> Regard t as a function (scalar field) defined once and for all in Newtonian spacetime

> $t = t(P)$ [P: spacetime point]
> Without it, spacetime could not be Newtonian, for "t" is every bit as intrinsic to the Newtonian spacetime as the metric "g" is to Lorentz spacetime. The layers of spacetime are the slices of constant t -the "space clices"- each of which has an identical geometrical structure: the old "absolute space".

The type of mathematical construction referred to by Weyl and Misner et al. as 'stratification' is commonly denoted in mathematics as foliation, and the manifolds to which it is applied are referred to as foliated manifolds.

For our purposes it is not necessary to go into all the details of a foliation (see Appendix). Roughly a three-dimensional foliation of a four-dimensional spacetime consists of a splitting up of the spacetime manifold into a family of three-dimensional submanifolds, called leaves, which are all disjoint. Thus the manifold is cut, so to speak, in slices so that each slice represents space at a certain moment of time (see fig. 12). These slices are referred to as strata by Weyl and as space-slices by Misner et al. By so foliating the spacetime manifold, we obtain a one-parameter family of leaves L_t (t: element from R); the order in which these leaves are arranged by the parameter t, correponds to the temporal order of the events lying in those leaves. Thus a foliation transforms a spacetime manifold into an infinitude of space manifolds ordered by a parameter which represents time.

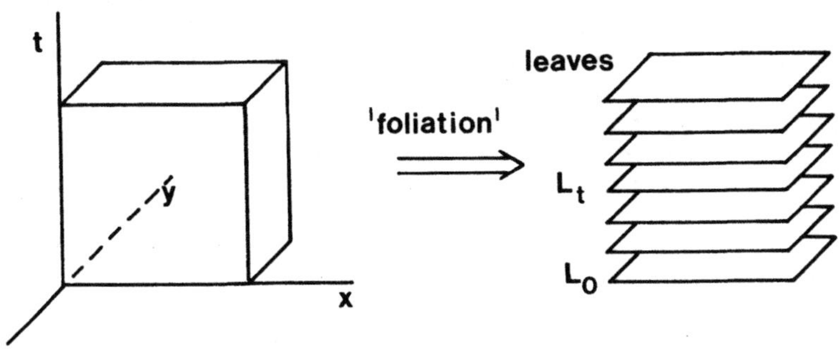

fig. 12
Pictorial representation of a foliation of spacetime.

Not all spacetime manifolds can so be foliated and thus, it is not the case that any arbitrary spacetime manifold allows for the reconstruction of Newtonian absolute space and time. The spacetime must satisfy certain conditions (see Appendix). In the following I shall assume that M is homeomorphic to E^4; since this is an open manifold, it allows a foliation of the type described above.

By endowing each one of the infinitude of spaces with the same geometrical structure, namely Euclidean geometry, and by assigning a fixed metric to the parameter used for labelling the leaves - in such way that differences in parameter values correspond to differences in absolute time for the events in the leaves - we come closer to a reconstruction of absolute space and time from Newtonian spacetime.

Here it should be stressed that the foliation of the spacetime manifold and the spatial and temporal metrics are to be regarded as part of the 'absolute' structure of Newtonian spacetime. (Misner et al. refer to the stratification as 'intrinsic' and as 'given once and for all'; Weyl describes it as 'objectively sinificant'.) This means that the spacetime manifold is foliated into spaces and time in the same way for each observer, regardless of his state of motion.

The determination of the metrical structure within each space (leaf) and of the metrical structure of the parameter, is not yet sufficient. The connection between the different spaces still has to be given. This can be done by fixing the worldlines of the 'points of absolute space', that is by giving for each point of a certain leaf the corresponding points in each of the other leaves (more precisely, one must specify a three-parameter congruence of curves [Earman 1970, p. 276-277]). The interpretation of these worldlines as geodesics in spacetime determines the connection coefficients [Misner 1973, p. 253]. The resulting structure of the spacetime manifold is all that is needed to reformulate Newtonian physics in the spacetime approach.

Two different paths can now be followed. In the first, Newtonian spacetime is taken to be flat; not only each space (leaf) is flat, but their connection is such that the whole spacetime is flat (vanishing connection coefficients). In the second, Newtonian spacetime is taken to be curved. In the former case, the gravitational force cannot be geometrised but a separate gravitational potential has to be introduced. In the latter, the effects of gravitation on matter can be incorporated into the geometric structure of Newtonian spacetime, analogously to the procedure followed by Einstein in the general theory of relativity [Misner 1973, Ch. 12].

In the following I shall only consider flat Newtonian spacetime; its structure is rather complicated because it does not possess a non-singular metric for spacetime intervals, as in relativistic spacetimes [Earman 1970, Stein 1967]. In fact, two different metrics, one for spatial and one for temporal intervals, are necessary. This means that in a Newtonian spacetime, the unification of space and time into a single spacetime is only partially successful.

The structure of flat Newtonian spacetime is that of a fiber bundle (10). For the description of flat Newtonian spacetime in terms of fiber bundles there are two possibilities:

a) Time is chosen as the base space and space as the fiber, as represented schematically in fig. 13a. In this case, we encounter an unique time but a multiplicity of spaces. This is the situation corresponding to the kind of foliation described above.

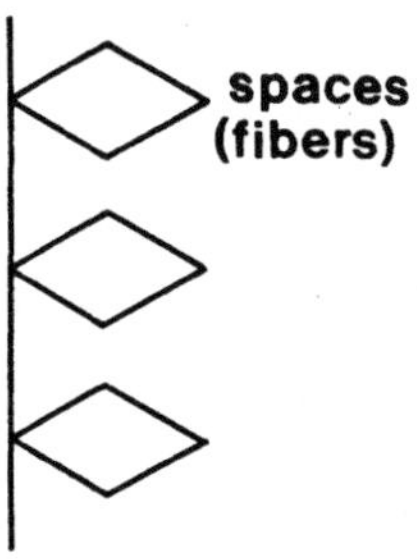

fig. 13a

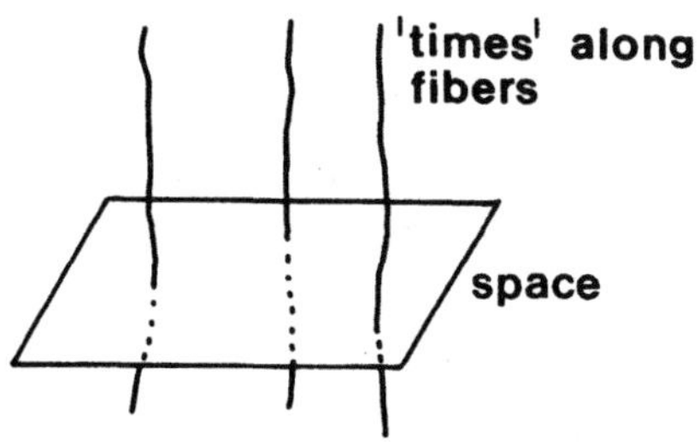

fig. 13b

b) Space is chosen as the base space and time as the fiber (see fig. 13b). Here we have an unique space and a multiplicity of times along each fiber (interpreted as a worldline). This approach corresponds to a one-dimensional foliation of spacetime: the leaves form a three-parameter family of one-dimensional submanifolds (the fibers) of the spacetime. The 'points' of the three-dimensional parameter space correspond to points of space.

Obviously, the first fiber bundle structure is more appropriate for the description of Newtonian spacetime, since absolute time has direct physical significance, whereas from a dynamical point of view, the objectivity of absolute space partly vanishes by virtue of Galilean relativity.

I shall now turn to an exposition of the role of parameter and coordinate time in the description of motion in Newtonian spacetime.

Motion in Newtonian spacetime.

Absolute time in Newtonian spacetime is a submanifold of spacetime and is treated as a fourth coordinate. The motion of a particle in Newtonian spacetime is again represented by a parametrised curve:

$$\begin{array}{llcl} & \text{Parameter: ?} & & \text{Newtonian spacetime} \\ C: & E & \longrightarrow & E^4 \\ & \tau & \longrightarrow & m(\tau) = (x(\tau), y(\tau), z(\tau), t(\tau)) \end{array}$$

We are now, of course, confronted with the problem of finding a physically acceptable interpretation of the parameter τ. Within a physical context, this parameter and the allowable class of reparametrisations must be defined in an objective way. Being part of the manifold in which motion is described, and functioning as a coordinate, absolute time t as such cannot function as the parameter, although it meets the requirement of having physical significance. However, if the worldlines are parametrised so that the parameter values correspond to the readings of a comoving clock, then the result is that the values of the parameter τ are equal to the values of coordinate time t along the worldline (or differing by a linear transformation). Havas, in his reformulation of Newtonian mechanics in the spacetime formalism, notes this point, but immediately adds that absolute time as a parameter τ must be distinguished from absolute time as a coordinate t [1964, p. 943] (11):

> Clearly τ [the parameter] differs from $z^0 = \pm t$ the coordinate at most by an additive constant. However, we prefer to distinguish the two quan-

tities notationally. First, this serves to emphasize that τ is a parameter on the world line rather than a coordinate (and thus could have been chosen differently from t), second, this renders the formulas of Newtonian and special relativistic mechanics much more symmetric..

This choice of parameter time τ meets the requirement of objectivity because differences in absolute time are objective within the Newtonian framework. The 'price' to be paid is that a double role must be attributed to Newton's absolute time in these four-dimensional reformulations: on the one hand, absolute time is treated as a submanifold of Newtonian spacetime and therefore functions as a coordinate, whereas on the other hand, it also functions as a parameter for the description of motion; and insofar absolute time is treated as a parameter, it is independent from the manifold in which motion is described.

The fact that parameter time is equal in value to coordinate time should not distract from the different geometrical role of each of them; each plays a fundamentally different role in the description of motion. Absolute time can fulfill this double role in the four-dimensional reformulations of Newtonian mechanics because of the specific behaviour of clocks within the Newtonian framework: once ideal clocks are synchronised, they stay synchronous forever regardless of their state of motion. In relativity theory this is no longer the case; clocks behave in a more complicated way and therefore, the relation between parameter and coordinate time becomes also more complicated.

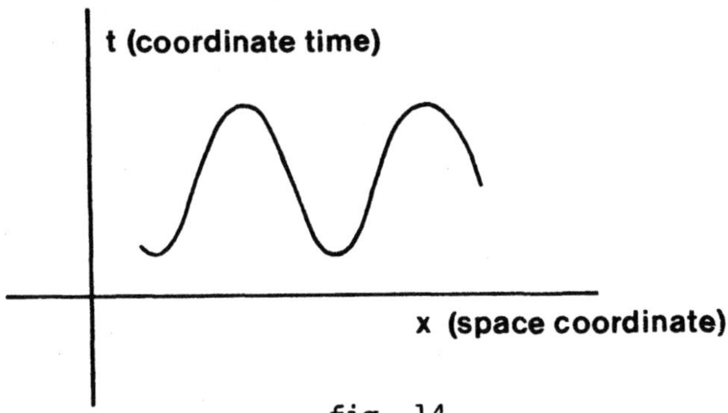

fig. 14

Since Newtonian spacetime is not a metric spacetime, the parameter cannot be defined in terms of the arc-length along a trajectory as is usually done in relativity theory. It is however possible to parametrise the geodesics, corresponding to particles in uniform motion, in a natural way with the help of the affine parameter, which is taken to be equal in value to absolute time [Trautman, in Hoffman 1966, p. 416]. But this method for parametrising trajectories does not work for curved worldlines.

Not every arbitrary curve in Newtonian spacetime represents the trajectory of a physically realisable motion. To avoid the awkwardness of one particle being at different places at the same moment of absolute time, only those curves belong to the class of trajectories which satisfy the following 'bijection condition': the coordinate time function t must be a bijection from the points of the curve to the reals (or in Weyl's terminology (p 70): "any fiber intersects a stratum in but a single worldpoint"). Therefore worldlines as given in fig. 14 must be excluded.

II.4 Parameter and coordinate time in Einsteinian physics

The association of the concept of spacetime with the theory of relativity has become so common in physics that both seem to form an inseparable pair. Nothing however is less true. The original formulation of the special theory of relativity was given by Einstein [1905] in the space and time language. Shortly after Einstein's first paper on relativity theory, Minkowski [1908] showed that this theory could be very elegantly rephrased by adding time as a fourth coordinate to the three space coordinates into a spacetime, now known as 'Minkowski spacetime'. In Einstein's original paper, Zur Elektrodynamik bewegter Körper, there is no reference at all to time as a coordinate or to the four-dimensional formalism. Therefore just as we spoke of the reformulations of Newtonian mechanics into the spacetime formalism, we can speak of the reformulation of special relativity theory into the spacetime formalism. But the specific structures of the corresponding spacetimes differ in many respects, reflecting the different concepts of space and time in respectively classical and relativistic physics. In the following, I shall restrict myself mainly to the special theory of relativity.

II.4.1 Relativistic space and time: proper time as parameter.

Since the notion of absolute space is abandoned within relativity theory, the motion of a particle in space cannot be described in an observer independent way; it is always necessary to specify the observer O (the reference frame) in relation to which a particle is moving. Relative to this observer, the motion of a particle is again given by a parametrised curve in space. For parametrising these trajectories, O has in principle two choices: either the trajectories are parametrised by time as it is measured by a clock which is at rest in space, that is by time t as measured by the observer O, or they are parametrised by time as measured by clocks moving along the trajectories, i.e. by proper time τ. Whereas in classical physics, there would be no difference between both procedures, in relativistic physics there is, since the rule that clocks, after being synchronised, remain synchronous is not valid. The first procedure is not appropriate because the resulting parametrisation of worldlines is not observer independent, whereas the parametrisation by proper time is observer independent.

The foregoing leads to some significant differences between the description of motion in Newtonian space and time and in relativistic space and time. In the first place, it becomes necessary to indicate the observer O or reference frame in relation to which motion takes place, whereas for Newton, this reference frame was uniquely determined, namely absolute space. This means that it is not possible to specify the manifold of scheme 1 in an observer independent way.

Secondly, it is not possible to parametrise the trajectories of different particles in the same coherent way as in the space and time formulation of Newtonian mechanics, where points on different trajectories with the same value of parameter time correspond to simultaneous events. Generally, points on different trajectories with the same proper time value (parameter value) are not simultaneous relatively to observer O; parameter time in relativity theory has lost its universal character.

Finally, in Newtonian space any continuous differentiable curve corresponds to a physically realisable motion of a particle. In relativity theory, however, the velocity of massive particles is always smaller than c and therefore only those curves for which at each point the velocity relative to O is smaller than c, represent physically realisable motions of massive particles. In other words, if the curves are parametrised by time t as measured by O, and not by proper time τ, than only those curves for which at each point $v = ds/dt$ ($ds = \sqrt{(dx^2 + dy^2 + dz^2)}$) is smaller than c, are trajectories of massive particles.

The two procedures for parametrising worldlines, mentioned above, are not independent from each other. For a particle in uniform motion, the relation between both is given by the famous time dilatation formula (which is a special case of the Lorentz-transformations):

$$\tau = t\sqrt{1 - (v/c)^2}$$

For non uniform motion the reparametrisation of a curve C from time t as measured by O to proper time τ is given by:

$$\tau = \int_C \sqrt{1 - (v/c)^2}\, dt$$

This formula also expresses the relation between the parameter times of two different trajectories, viz. the parameter time t of the trajectory of O) which consists of just one point in space because O is at rest) and the parameter τ of trajectory C. This shows that for the comparison of the parameter times of different trajectories, the relative velocity of the corresponding particles plays an important role. Thus in spite of the fact that parameter time is defined such that it is independent of the manifold, the correlation between the parameters of different trajectories cannot be given independently of the manifold because it is necessary to refer to the relative velocity of the particles involved. At this point, the intimate relationship between space and time emerges within the space and time formulation of the special theory of relativity.

II.4.2 Relativistic spacetime: proper time and coordinate time.

The formulation of the special theory of relativity in the spacetime formalism, takes as its point of departure Minkowski spacetime which, in contrast to Newtonian spacetime, possesses a spacetime metric. The differential of the spacetime distance is given, in suitable coordinates, by

$$ds = \sqrt{((cdt)^2 - dx^2 - dy^2 - dz^2)}.$$

As is well known, this quantity is left invariant under Lorentz transformations. Thus, only the spacetime distance between two events has objective physical significance [Minkowski in Einstein 1952, p. 75]:

> Henceforth space by itself, and time by itself, are doomed to fade away into mere shadows, and only a kind of union of the two will preserve an independent reality.

Neither the spatial nor the temporal distance between events is invariant under Lorentz transformations.

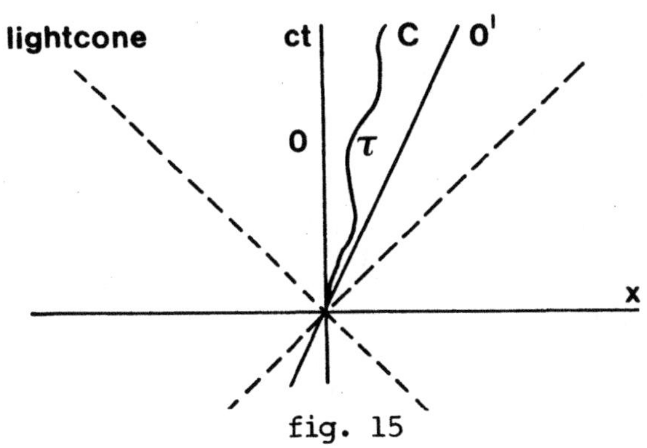

fig. 15

As in the space and time formulation, the abandonment of absolute time and the relativity of uniform motion demand the arbitrary selection of an inertial observer O who is at rest in the corresponding Minkowski spacetime. His trajectory coincides with the time coordinate axis. The trajectory of a particle in this Minkowski spacetime is given by:

$$C \qquad \tau \longrightarrow m(\tau) = (x(\tau), y(\tau), z(\tau), t(\tau)).$$

The trajectories of observers in uniform motion relative to O are given by straight lines (see fig. 15).

In Newtonian spacetime, any curve satisfying the bijection condition represented a physically realisable motion. In Minkowski spacetime, however, only those curves for which at each point the tangent vector lies within the lightcone, can correspond to the trajectories of massive particles, since for these alone, the velocity of the particles relative to O, or to any other inertial observer O', is smaller than the velocity of light.

The proper time τ along a curve C is given by the spacetime distance along C:

$$\tau = \int_C ds$$

More in particular:

$$\tau = \int_C ds = \int_C \sqrt{1 - (v/c)^2}\, dt$$

As was to be expected, this expression is equivalent to the one given at page 78, but now it has a different meaning. It gives the relation between parameter time (proper time) and coordinate time of Minkowski spacetime.

In general, differences in coordinate time t are not equal to differences in parameter time τ for points on a trajectory. But besides being unequal in numerical value, parameter time τ and coordinate time t also differ strongly with regard to their behaviour under Lorentz transformations. Parameter time is left invariant under those transformations and therefore has direct physical significance; this is not the case for coordinate time.

II.4.3 Parameter and coordinate time in the general theory of relativity.

Whereas the special theory of relativity can be formulated within both the space and time and the spacetime formalism, the situation with regard to the general theory of relativity is more complicated. Einstein's original formulation of this theory makes use of the spacetime formalism, and a reformulation in the space and time formalism has never been given. The question is whether such a reformulation is possible. According to Graves [1971, p. 137]:

> it is still possible to state the principles of SR [the special theory of relativity] and give their physical significance without using the formalism of four-vectors and tensors; it is NOT POSSIBLE to do this in GR [the general theory of relativity].

Graves claims that the principle of "general covariance commits physics to the use of tensors... for the representation of all its fundamental quantities" and that therefore the spacetime formalism has become indispensable for the formulation of the principles of general relativity [ibidem]. Einstein is less outspoken in his evaluation of the role of the spacetime formalism with respect to the general theory. For him there is no doubt that the "four-dimensional mode of consideration of the 'world' is natural on the theory of relativity" [Einstein 1954, p. 56]; but he is less categorical than Graves with regard to the question whether the four-dimensional formalism is an indispensable tool for formulating general relativity for he remarks that "without it [the four-dimensional formalism] the general theory of relativity....would perhaps have got no farther than its long clothes" [op. cit. p.57].

Surely, any attempt to reformulate general relativity in the space and time formalism would necessitate a revision of the standard scheme for the description of motion (p 59), since the structure of space would have to vary in time. In other words, the structure of the manifold, in which motion is represented by a parametrised curve, would also become a function of the parameter. So far it was tacitly assumed that the structure of the manifold is itself parameter (time) independent. This assumption would have to be abandoned; it would not be possible to choose the parameter and the manifold of scheme 1 so that they are independent of each other. Therefore, within the space and time formulation of the general theory of relativity, the intimate relationship between space and time would already become manifest at this very fundamental level. However, it is an open question whether this adaptation of the standard scheme for the description of motion would allow for the reformulation of general relativity in the space and time formalism.

The description of motion in general relativity proceeds along the same lines as in the spacetime formulation of the special theory of relativity, except for the fact that flat Minkowski spacetime is replaced by a curved Riemannian spacetime. This does not affect the role of parameter and coordinate time, the more so since locally curved Riemannian spacetime looks like flat Minkowski spacetime. A problem arises if the curved spacetime possesses a cyclic coordinate time. In that case, the events on a closed worldline have a cyclic temporal order relative to coordinate time and a linear temporal order relative to parameter time (because parameter time is homeomorphic to the standard open interval I). (One obvious way to remove this conflict between both orderings is to adjust the structure of the parameter space in the standard scheme so that it becomes homeomorphic to S^1).

II.5 The geometrical nature of time in quantum mechanics

In his article "Ueber den anschaulichen Inhalt der quantentheoretischen Kinematik und Mechanik", Heisenberg [1927] discusses the notion of the 1S-trajectory (orbit) of an electron in a hydrogen atom. After having remarked that, in order to determine the different places corresponding to the 1S-trajectory of the electron, the hydrogen atom would have to be illuminated with light of a wavelength much smaller than 10^{-8} cm, he writes (12):

> One single quantum however of such light is sufficient to throw the electron completely out of his 'trajectory' (therefore always only one single point in space of such a trajectory can be defined); the word 'trajectory' has thus here no sensible meaning ('keine vernünftigen Sinn').

Thus at the microscopic level, we cannot trace by observation the motion of a particle through space because the interaction involved in the process of observation will disturb the motion of the particle in an unpredictable way. This is formally expressed in Heisenberg's uncertainty relation:

$$\Delta x \Delta p_x \geq \hbar/2$$

If the variation in the position of the electron is reduced to zero, i.e., if the position of the particle is exactly known, then the variation in the momentum is infinite, i.e., there is no information about the momentum of the particle. In terms of spacetime diagrams, this means that exact knowledge of a point of the trajectory of the particle implies total ignorance about the tangent vector of the trajectory at that point; or conversely, exact knowledge of the tangent vector implies total ignorance about the point of the trajectory.

At first sight, Heisenberg's considerations seem to put into question the validity of the standard scheme for the description of motion in quantum mechanics. If space is taken to be the manifold in which motion is described, then the notion of a trajectory is meaningless, and therefore the motion of a particle cannot be represented by a parametrised curve in space. A closer look reveals that this is not necessarily so. By making an appropriate choice for the manifold M, it is possible to describe the evolution of a quantum mechanical system by a parametrised curve. But the price is rather high; the manifold in question is a highly abstract space, viz. a Hilbert space. The evolution of a quantum mechanical system can be represented by a parametrised curve in Hilbert space:

	Time		Hilbert space
C:	E	----->	H
	t	----->	$x_1(t),\ldots\ldots,x_n(t),\ldots$

Time again plays the role of parameter, which is in agreement with the fact that time enters as a real parameter in the Schrödinger equation [von Neumann 1955, p. 354].

A comparison of the role of space and time in the Schrödinger formalism of quantum mechanics shows that they are quite different in nature. Whereas the position in space, as well as all other observables, are represented by Hermitian operators, time is represented by a parameter. If Dirac's distinction [Dirac 1926] between c-numbers (c-numbers are the variables of classical physics satisfying the commutatitve law) and q-numbers (q-numbers are the quantum mechanical variables which do not always satisfy the commutative law) is used, the fundamental difference between space and time becomes immediately clear: whereas time t is a c-number, the space coordinates x_i are q-numbers. The fact that all observables are represented by Hermitian operators except for time which is merely a parameter, is often experienced as an awkward dissymmetry in quantum mechanics; but none of the proposals to remove this dissymmetry by replacing time as a parameter by time as an operator has yet proven to be successful (13).

In relativistic quantum mechanics, the picture of the geometrical nature of time becomes even more complicated. The Dirac equation describing the evolution of a one-particle quantum mechanical system is invariant under Lorentz transformations [e.g. Sakurai 1973, ch. 3]. This suggests that time t is treated as a fourth coordinate. But the same variable is also used as parameter time since the evolution of the system is given relatively to the same variable t. Thus, it appears at least that parameter and coordinate time are not distinguished notationally. Moreover, the interpretation of the time variable is rather problematic. If time t is indeed treated as a coordinate, then one would expect a sensible definition and interpretation of the expectation value of time, $\langle t \rangle$, analogously to the definition and interpretation of the expectation value of the space coordinates, $\langle x_i \rangle$. But a meaningful interpretation of the expectation value of time t is lacking. To conclude this short excursion into quantum physics, the geometrical nature of time in quantum mechanics, both non-relativistic and relativistic, is rather obscure.

II.6 Translation of the space and time into the spacetime formalism and vice versa

I shall now briefly examine the possibility of the passage from the space and time to the spacetime formalism and conversely, from the spacetime to the space and time formalism.

In the first place, it is necessary to establish a correspondence between 'material particles' located in space and 'events' located in spacetime. This can easily be achieved with the notion of genidentity which is the material counterpart of the logical notion of identity [Reichenbach 1971, p. 38]. The equivalence classes generated by the application of the relation 'being genidentical with' to the events of spacetime, represent material objects. Each equivalence class contains the events belonging to the history of a particle. Conversely, an event corresponds to a material particle located at a definite place in space at a definite moment of time.

Given these correspondence rules for material particles and events, the translation from the space and time to the spacetime formalism does not pose any serious problems. It can be achieved by simply taking as the spacetime manifold the direct product of the original space and time and by assigning to the new parameter for the description of motion in this spacetime the same value as coordinate time (as was the case in the four-dimensional reformulations of Newtonian mechanics).

The transition from the spacetime to the space and time formalism is somewhat more complicated. How is it possible to recover the notions of space and of time starting from a spacetime manifold? As was already seen in the case of the reconstruction of Newtonian absolute space and time starting from a Newtonian spacetime, this can be achieved by foliating a spacetime manifold, which leads to an infinitude of space manifolds ordered by a parameter which represents time. This infinitude of spaces can be reduced to an unique space by a suitable identification of points from different spaces. The result is an unique space which functions in the space and time formalism as the manifold in which motion is described.

The conditions under which a spacetime manifold can be foliated so that it is finally split up into a three-dimensional space and a one-dimensional time, depend among other things upon the number of dimensions of the manifold and upon its topological structure (see Appendix).

What happens to the distinction between parameter and coordinate time in passing from the spacetime description via a foliation to the space and time description? One starts from the two notions of parameter and coordinate time, and only the notion of parameter time is retained. The effect of foliating the spacetime manifold is that the distinction between parameter and coordinate time becomes obliterated because parameter time becomes identified with coordinate time. To illustrate this, consider a parametrised curve $c(\tau)$ in a spacetime manifold, satisfying the bijection condition (see p 76):

$$c(\tau) = (x(\tau), y(\tau), z(\tau), t(\tau)).$$

Because of the bijection condition, the points of the curve can also be parametrised by the values of coordinate time which means that τ can be choosen equal to t:

$$c(t) = (x(t), y(t), z(t), t).$$

Now, if coordinate time is reinterpreted as a parameter, that is if we pass throuh a foliation from the spacetime to the space and time formalism, then $(x(t), y(t), z(t))$ represents a curve in three-dimensional space parametrised by time t. In other words, the distinction between parameter and coordinate time in the spacetime formalism disappears because the variable t corresponds not only to coordinate time but also to parameter time.

The foregoing shows the importance of the bijection condition with respect to the problem of translating the spacetime formalism into the space and time formalism. Without this bijection condition the reinterpretation of coordinate time as parameter time would not be possible. Particles travelling back and forth relatively to coordinate time, as for instance occur in the Feynman-Stueckelberg interpretation of a positron as an electron travelling back in (coordinate) time, must be excluded in the spacetime formalism. If such curves are admitted as trajectories of real particles, then immediately the possibility of parametrising these trajectories by coordinate time is destroyed; then it also becomes impossible to identify coordinate time with parameter time and to pass from the spacetime to the space and time formalism.

II.7 The geometrical nature of time and dynamics

In this section, I shall briefly consider the role of parameter and coordinate time from the point of view of dynamics. To my knowledge, this topic has only received attention in the work of Lautman and Scheurer. Both attach great importance to the distinction between parameter and coordinate time but reach divergent conclusions with regard to its role in dynamics. Here, I shall only present a schematic outline of their ideas.

In "Le problème du temps", Albert Lautman [1977] analyses what he calls the 'sensible' properties of time and introduces the distinction between the geometrical properties of time, which are related to the notions of dimension and orientation, and the dynamical properties which are related to the notion of evolution [op. cit. p. 257]. This distinction runs parallel to the distinction between parameter time and coordinate time since the dynamical aspects of time are related to the use of time as a parameter (of evolution) and the geometrical aspects to the use of time as a coordinate [op. cit. p. 272].

Lautman's starting point for introducing the distinction between the geometrical and the dynamical aspects of time is not the standard scheme for the description of motion, but mathematics. According to Lautman, this double aspect of time can be deduced a priori, that is, within the realm of pure mathematics. In a nutshell, Lautman claims that the distinction between the geometrical and the dynamical aspects of time has its origin in the distinction between a system of first order partial differential equations and the corresponding 'characteristic' equations as defined by Monge [op. cit. p. 266 ff]. The derivation with regard to time as a coordinate appears in the system of partial differential equations itself, whereas the derivation with regard to time as a parameter is found in the characteristic equations.

Of special interest is Lautman's analysis of the consequences of treating time as a parameter and as a coordinate at the dynamical level. He claims that within classical mechanics, as opposed to quantum mechanics, there is a rigorous equivalence between the treatment of time as a parameter and as a coordinate. According to Lautman, this equivalence follows from the fact that the Hamiltonian principle of dynamics, from which all of classical mechanics can be derived, can be formulated in two different, but equivalent ways, namely with time as a parameter and with time as a coordinate [op. cit. p. 259-261]. In other words, as far as classical mechanics is concerned, a translation of the spacetime formulation of this theory into the space and time formalism (and vice versa) would not only be possible at the level of kinematics, but also at the level of dynamics. Thus, Lautman reaches the conclusion that, within classical mechanics, it does not matter whether time is treated as a parameter or as a coordinate.

Scheurer [1973,1977,1982], however, presents a different view on the role of parameter time and coordinate time in dynamics. In his opinion, it is possible to 'derevolutionise' the special theory of relativity, i.e., to derive both classical and relativistic mechanics from a common dynamical scheme based upon the use of the theory of differentiable manifolds, if due consideration is given to the geometrical nature of time. Within this scheme, the treatment of time as a parameter leads to classical Newtonian mechanics whereas the treatment of time as a coordinate leads to relativistic Einsteinian mechanics. Thus, there is no equivalence at all between the treatment of time as a parameter and as a coordinate within this dynamical scheme.

According to Scheurer, there is a fundamental ambiguity in classical mechanics with regard to the role of time. This he illustrates with the example of the 'convective derivative' [Scheurer 1982, p. 218]:

$$\frac{d}{dt} = \frac{dx^i}{dt}\frac{\delta}{\delta x^i} + \frac{\delta}{\delta t}$$

In this expression, the time variable t plays a double role: in terms of the total derivative with regard to t, time functions as the parameter of evolution whereas the partial derivative with regard to time t refers to the use of time as a coordinate. To remove this ambiguity, it is necessary to code time into two distinct mathematical entities, namely parameter and coordinate time.

This makes possible, according to Scheurer, a structural reinterpretation of the special theory of relativity. Within the present context, one of the most interesting features of this structural reinterpretation is that, on the level of dynamics, the distinction between parameter and coordinate time entails the distinction between proper mass and mass [Scheurer 1977, p. 163] (14):

> In fact, time t having become a coordinate, it was necessary for Einstein to introduce a new parameter on the manifold, viz. proper time, in order to treat motions in spacetime as parametrised curves. But at the level of dynamics, he was at the same time also forced to introduce a new mass - the rest mass m_o - which should stand in the relation to proper time as the mass m to time [coordinate time t].

Starting from this relationship, Scheurer proposes a reformulation of both Newtonian and Einsteinian mechanics in terms of an unique dynamical structure by means of the so-called Newton-Einstein axiom:

$$\frac{dx^i}{p^i} = \frac{dt}{m} \overset{\text{déf}}{=} \frac{dx^4}{p^4} = \frac{d\tau}{m_o}$$

which, he claims, is valid for any differentiable manifold in which motion is represented by a parametrised curve [1977, p. 163]. From this axiom, the equivalence between mass and energy can easily be derived [Scheurer 1977, p. 165, 1982, p. 223].

I shall not go further into these different views on the relevance of the distinction between parameter and coordinate time for the structure of dynamics. For more details, I refer the reader to the relevant literature.

II.8 Parameter versus coordinate time in the study of time; some philosophical issues

There can be no doubt that within physics the spacetime formalism has proven to be more powerful than the space and time formalism; therefore the spacetime description of physical reality is generally considered to be more fundamental than the space and time description. In this final section, I shall consider the question whether the spacetime formalism, in particular the distinction between parameter time and coordinate time, is of any importance for the study of time. On this matter, opinions have been divided.

In his influential work The philosophy of space and time, Reichenbach writes the following [1957, p. 110-112]:

> Whereas the conception of space and time as a four-dimensional manifold has been very fruitful for mathematical physics, its effect in the field of epistemology has been only to confuse the issue....
> Our schematization of time as a fourth dimension therefore does not imply any changes in the conception of time....
> The properties of time which the theory of relativity has discovered have nothing to do with its treatment as fourth dimension. This procedure was already possible in classical mechanics, where it was frequently used. However, according to the theory of relativity the four-dimensional manifold is of a new type; it obeys laws different from those of classical theory.

Reichenbach even warns us to be on our guard because "calling time the fourth dimension gives it an air of mystery" [op. cit. 110], and such expressions can easily be misunderstood or mislead us in our analysis of the concept of time. The properties of time, as exhibited by the theory of relativity, cannot be traced back to time being a coordinate [op. cit. 112]:

> To demonstrate these changes in the conception of time we need not employ mathematical con-

> siderations. We can remain within the perceptual experience of time and develop everything the theory of relativity teaches about time.

Clearly, for Reichenbach the spacetime language does not constitute a more powerful language for dealing with the problem of physical time.

An outspoken advocate of the use of the spacetime formalism in the study of time is Earman. In a paper with the rather provoking title "Space-time, or how to solve philosophical problems and dissolve philosophical muddles without really trying", Earman [1970] tries to show how long-standing philosophical issues about space and time can be fruitfully discussed if the notion of spacetime is taken seriously. His central claim is that the "failure to take space-time seriously has resulted in much philosophical debate about an ill-formulated issue" [1970, p. 261]. At the end of his article, Earman writes [p. 274]:

> But I think enough has been said to show that the failure of the philosophers fully to appreciate the fact that space-time is the basic spatio-temporal entity has led to an impoverishment of that area of philosophy usually called the philosophy of space and time. In the first place, it has been responsible for the fact that philosophers have overlooked the complex and subtle relations between space and time, that is, the spatial and temporal aspects of space-time. Secondly, it has led to a neglect of a powerful method for attacking philosophical problems about space and time: take some thesis allegedly about space or time simpliciter and translate it into a thesis about the structure of space-time; and then bring to bear all the known mathematical results and physical evidence about space-time structure. At the very least, this method produces some clarification, e.g., by reducing the original problem to a precise mathematical or physical question; at best, it can lead to a quick solution of the original problem or a dissolution of the original muddle, and it can also suggest new twists to old questions.

Thus, according to Earman the best recipe for solving problems (or muddles) about time consists in reformulating or translating these problems into the spacetime language.

A cursory look at the recent literature concerning the study of time is sufficient to conclude that the balance is definitely in favour of Earman's point of view. Despite Reichenbach's warning against "calling time the fourth dimension" it becomes standard practice in the study of time to exploit the fact that in the spacetime formalism time functions as the fourth coordinate of the spacetime manifold and often the Earman recipe is successful, although it does not turn out to be the panacea he suggests it is.

In my opinion however, some of the philosophical consequences of the use of the spacetime formalism for the description of physical reality, as well as for the study of time, have been misunderstood. Inspired by the famous passage of Weyl, quoted at page 61, both scientists and philosophers have claimed that relativity theory leads to a completely static conception of the physical world in which the future is determined because all events are given once and for all in the spacetime manifold. Moreover, the conclusion is drawn that time becomes spatialised because relativity theory treats it in the same way as it treats space.

My intention is to show that such conclusions are unfounded because of the following reasons. In the first place, they are based upon the tacit assumption that relativity theory and the spacetime formalism are an inseparable pair; it is taken for granted that relativity theory implies a spacetime description of physical reality. In the foregoing, we have seen that, as far as the special theory of relativity is concerned, this is not the case. Of course there can be no doubt that, as Einstein [1954, p. 51] remarked, the "four-dimensional mode of consideration of the 'world' is natural on the theory of relativity", but an analysis of the philosophical consequences of relativity theory as such for the physicist's conception of the world, is best served by abstracting from the peculiarities of the language in which the theory is expressed. Therefore, the by now non-standard space and time formulation of the special theory of relativity deserves also be taken into consideration.

Secondly, even if, for the sake of the argument, we accept that special relativity implies the use of the spacetime language (because this language leads to the most 'natural' or 'elegant' formulation of this theory), then it would still be questionable whether such conclusions are valid. The reason for doubt lies in the fact that the role of time in the spacetime formalism has been generally misunderstood because the dual geometrical nature of time was not taken into account. Time is considered to be only a coordinate whereas in fact it functions as both coordinate and parameter.

Keeping in mind i) that there is no intrinsic connection between the special theory of relativity and the spacetime formalism, and ii) the dual geometrical nature of time in the spacetime formalism, I shall now examine in more detail the claims that special relativity theory implies or leads to 1) a 'spatialisation' of time, 2) a static world picture, and 3) determinism.

1) Spatialisation of time.

The expression 'spatialisation' of time refers to a conception of time based on the model of space. A spatialisation of time lies at the basis of the Newtonian conception of time as a straight line along which the now or the present moves in the direction of the future. In this model, the instants of time correspond to geometric points. Assuming, as is quite common among physicists (see Epilogue), that the motion of the now along this line (that is the flow of time) is a subjective phenomenon and that therefore it is not part of the structure of physical time, the Newtonian conception of time degenerates into a straight line; time becomes isomorphic to the Euclidean line (15).

It is a widespread idea that relativity theory strengthened this tendency to spatialise time i) by treating time on a par with space, i.e. as a coordinate, in the Lorentz transformations which revealed the intimate relationship between space and time, and ii) through the unification of space and time in a four-dimensional spacetime by means of the universal constant of homogenisation c which allows time intervals to be measured in units of length and vice versa. Because of this unification, the specific character of time appears to be lost almost completely.

It cannot be denied that relativity theory disclosed

an intimate relationship between space and time. It is however by no means necessary to interpret this relationship as implying a spatialisation of time in the sense that time is treated on the same footing as space. The original space and time formulation of special relativity clearly shows that this theory does not require time to be treated as a fourth coordinate. But even within the spacetime formulation of relativity theory, the spatialisation of time is of a restricted nature because it pertains only to coordinate time and not to parameter time. Coordinate time is indeed an almost completely spatialised time for, structurally, a space coordinate cannot be distinguished from the time coordinate in the spacetime manifold. Clearly, coordinate time is in a special position because it differs in sign from the space coordinates in the spacetime metric, but this is not due to a difference in the internal structure of the space and time coordinates. This special feature of coordinate time is related to the fact that in passing from the spacetime to the space and time formalism, the time coordinate alone can be reinterpreted as a parameter for the parametrisation of worldlines. Coordinate time distinguishes itself from the space coordinates only by the possibility of its identification with parameter time.

The spatialisation of time in the spacetime formalism does not extend to parameter time. In fact there is no difference at all between parameter time in the space and time and in the spacetime formalism. Therefore, inasmuch as time plays a role distinct from space in the space and time formalism, it does so also, in terms of the notion of parameter time, in the spacetime formalism. Since parameter time plays a primary role in the description of the evolution of a system, which by definition takes place from the past toward the future, the distinction between past and future orientation is part of the structure of parameter time. In other words, both orientations of parameter time are not equivalent. This supplementary structure is precisely absent in both coordinate time and the space coordinates. There is no intrinsic difference between past and future orientation of coordinate time. Without such intrinsic difference, parameter time could not fulfill its primary function of labeling unambiguously the states in the evolution of a system.

In summary, neither special relativity as such nor the spacetime formalism in which this theory is usually formulated, leads to a spatialisation of time; within the spacetime formalism there is a spatialisation of coordinate time but not of parameter time.

2) Static world picture.
According to Weyl the theory of relativity has shown us the real, Parmenidean nature of the objective world: "the objective world simply is, it does not happen". The physical universe is a static 'block' universe in which there is no real change. Only with regard to a conscious observer "crawling upward along his life line" does it make sense to speak of change. The flow of time is therefore not an objective phenomenon, as in Newton's world picture, but is subjective or minddependent. The question to be considered here is whether indeed relativity theory or the spacetime formalism implies that the flow of time is minddependent (16)?

If we compare the space and time formulation of Newtonian physics with the space and time formulation of special relativity theory, there seems to be no reason at all to conclude that in the former the flow of time is objective and that in the latter it is subjective. Of course, there is a significant difference between both theories with regard to the flow of time, but this difference is not related to the issue of the objectivity or minddependency of the flow of time. Within Newtonian physics, the flow of time has a universal character because of absolute simultaneity. In other words, the sets of events that consecutively constitute the present are the same for every observer, regardless of his place or state of motion. Within relativity theory this is not true because simultaneity is observerdependent, and thus observers in different states of motion disagree about which events constitute the present. 'Observerdependent' in this context means 'dependent upon the state of motion of the observer'; thus, time flow looses its universal character, but this by itself does not imply that it also becomes minddependent.

Thus, the real issue is not whether relativity theory banishes change from our picture of the physical world, but whether the spacetime formalism necessarily involves a static worldview. Again, due consideration of the dual geometrical nature of time in that formalism, in particular of the role of parameter time, shows that this is not the case.

It can hardly be maintained that there is a flow of coordinate time and thus, that there is change or becoming with regard to coordinate time. Relatively to coordinate time, all worldlines and events are given. This is completely in line with our remarks about the spatialisation of coordinate time. But this does not exclude the possibility of coherently relating the notion of the flow of time to parameter time. In the space and time formulation of Newtonian physics, the increase of parameter time represents the objective flow of absolute time; for increasing values of parameter time, the distribution of the particles in space will be different, and therefore there is change and becoming with regard to parameter time. However, the same kind of reasoning, applied to parameter time in the spacetime formalism of relativity theory, leads to the conclusion that parameter time has an objective flow (but with the proviso that the flow of parameter time is not universal).

Taking into account the role of parameter time in the spacetime formalism, we may conclude that there is no difference between the space and time and the spacetime description of physical reality with regard to the question whether the flow of time is objective or not. In as much as change and time flow can be regarded to be 'real' within the space and time description of physical reality, they can also in the same sense be regarded as 'real' in the spacetime description. The introduction of the notion of coordinate time in the standard formulations of relativity theory does not by itself lead to a static conception of the physical world.

3) Determinism.

The problem of determinism has received considerable attention within discussions of the implications of the theory of relativity for the physicist's conception of nature. Often it is suggested that there is a close connection between relativity theory and determinism; Rietdijk [1966] even goes so far as to claim that the special theory of relativity rigorously implies determinism. The main problem in evaluating such claims lies in the meaning of the notion of determinism.

The most famous doctrine of determinism is Laplacean determinism which says that, given the state of the universe at any arbitrary moment, it is in principle possible to predict (or retrodict) the whole future (past). But the claim that relativistic mechanics implies Laplacean determinism is not a very interesting one because i) Laplacean determinism is not a specific feature of relativistic mechanics since classical mechanics is also deterministic in the Laplacean sense, and ii) Laplacean determinism is not inherent to the spacetime formalism; within a spacetime manifold, events are not necessarily functionally related so as to make possible the prediction (retrodiction) of the future (past). Of greater interest, at first sight, is the claim that the special theory of relativity implies that the future is 'determined' or 'fixed' because in a relativistic spacetime the past, present and future are given all together.

Clearly, this kind of determinism is not inherent to the special theory of relativity itself since this theory can also be formulated without the use of the spacetime formalism. So the question remains whether the spacetime formalism implies this kind of determinism. The problem with finding an answer is that it is very difficult to give a precise meaning to the notion of a 'fixed' future. A fixed future is supposed to be the opposite of an 'open' or 'alterable' future, that is, a future which allows for real possibilities. In contrast with Laplacean determinism, this kind of determinism does not imply that it is also possible to predict future events. Although future events are not necessarily predictible, they are nevertheless regarded to be determinate or, as Grünbaum [1971, p. 223] says, they are supposed to have an intrinsic attribute-specificity: they are what they are and could not be otherwise. In this conception, the future is not alterable, but merely hides our destiny or fate. Understood in this sense, however, the statement that the future is fixed becomes a triviality because, as Grünbaum [ibidem] remarks: "it is a fact of logic that what will be, will be". So, this form of determinism does not in any way restrict the nature of physical reality, and therefore the statement that relativity theory or the spacetime formalism implies that the future is fixed, becomes a true but empty statement.

Clearly, the whole issue about relativity theory and determinism hinges on the meaning of the notion of determinism. In my opinion, a physical interpretation of determinism that would endow the claim that relativity theory or the spacetime formalism implies determinism with some non-trivial content, has yet to be formulated.

In most discussions of the role of time in relativity theory, all attention is focussed on the treatment of time as a coordinate, while the role of time as a parameter stays in the dark. This gives time 'an air of mystery' and leads to a number of philosophical problems. In the foregoing, I have tried to show that there is nothing mysterious about treating time as a fourth coordinate in relativity theory provided that the dual geometrical nature of time is taken into account.

APPENDIX

A foliation is defined in the following way [Iyanaga 1977, p. 359]. All manifolds and maps are C^∞. Let M be a n-dimensional manifold. A codimension q foliation of M $(0 \leq q \leq n)$ is a family $F = \{L_\alpha;\ \alpha \in A\}$ consisting of arcwise connected subsets of M, called leaves, with the following properties:

i) $L_\alpha \cap L_{\alpha'} = \emptyset$ if $\alpha \neq \alpha'$,

ii) $\cup_{\alpha \in A} L_\alpha = M$,

iii) Every point of M has a local coordinate system (U, Ψ) such that, for each L_α, the arcwise connected components of $U \cap L_\alpha$ are described by x^{n-q+1}= constant,....., x^n= constant, where $x^1, x^2, \ldots, x^n$ denote the local coordinates in the system (U, Ψ). In particular, every leaf of F is an (n-q)-dimensional submanifold of M.
For instance, the totality of integral curves of a non-singular vector-field on a n-dimensional manifold M constitutes a codimension n-1 foliation of M. A foliation can also be given by a submersion: let Q be a q-dimensional manifold $(q \leq n)$ and let f: M ---> Q be a submersion; then f induces a codimension q foliation of M with the arcwise connected components being defined by

$f^{-1}(x)$, with x a point from Q. In order to admit a foliation, M has to satisfy certain conditions. These conditions depend upon the number of dimensions of the manifold itself, upon the codimension of the foliation and upon the topological structure of M. For instance, every open manifold admits a codimension 1 foliation, whereas, for a closed manifold M, a codimension 1 foliation is only possible if the Euler number of M is zero. Take M to be a four-dimensional spacetime manifold. A codimension 1 foliation of M leads to a family of leaves each of which represents space at a certain moment of time; time is represented by the indexing set. Not every spacetime admits such a foliation. Newtonian and Minkowski spacetime can be foliated in this way because they are open manifolds (both are homeomorphic to E^4). But for instance a spacetime M homeomorphic to S^4 does not admit such a foliation because the Euler number is unequal to zero (no non-singular vector-field exists on S^4).

NOTES

1 Certain types of change cannot be represented by a trajectory in some space, e.g. the 'discrete jump' of an electron from one energy level to another in Bohr's 1913 atomic theory.

2 For an underpinning of the differentiable structure of spacetime, see: J. Ehlers et al. "The geometry of free fall and light propagation" in O'Raifeartaigh [1972] and Woodhouse [1973].

3 That the number of dimensions of U is equal to the number of coordinate functions x_i, is a fact often taken for granted. However this is not self-evident, and was proven by Brouwer; he showed that the number of dimensions of Euclidean n-space (which is the ordinary space of n real variables) is exactly n [Hurewicz 1941, p. 4-5].

4 'Dual' here means interchanging domain and codomain of a function. It should not be confused with the 'duality' of sets and topological spaces [cfr. Choquet-Bruhat 1977, p. 10, 21].

5 Moreover, by choosing the identity morphism as the coordinate function, the difference between a point of the manifold and its coordinate also disappears, and thus x stands for both a point of the manifold and for its coordinate value.

6 Note that by interpreting the parameter in the standard scheme for the description of motion as time, parameter time becomes homeomorphic to I; within the spacetime description, this choice for a linear parameter time may clash with a cyclic coordinate time (see p 82).

7 The absoluteness of space does not by itself imply that space must be Euclidean [Earman 1970a, p. 290].

8 Galileo, Descartes and Newton introduced the notion of state of motion [Koyré 1965, p. 188 ff].

9 For the notion of stratification see also J. Ehlers, "The nature and structure of spacetime" in Mehra [1973].

10 Roughly, a fiber bundle structure is a generalisation of the more simple direct product structure of two spaces. For an elementary description of a fiber bundle and its relevance to the problem of space, time and spacetime, see Trautman, "Theory of gravitation" in Mehra [1973].

11 Havas gives a third reason which concerns the fact that, in a generally covariant formulation of Newtonian mechanics, parameter and coordinate time are in general not equal in value. This can be seen as follows: to achieve the most general covariant formulation of Newtonian mechanics, Havas has to admit the largest class of allowable coordinate transformations which leave absolute simultaneity intact. This class consists of all reparametrisations of coordinate time [op. cit. p. 953]. But a reparametrisation of coordinate time does not change the parametrisation of a worldline by its proper time; this parametrisation is given once and for all. Therefore, parameter and coordinate time are no longer equal in value in the most general covariant formulation of Newtonian mechanics [op. cit. p. 954, equation 118N].

12 Heisenberg [1927, p. 172]; the translation is mine.

13 See, for instance Rankin [1965] and Prigogine [1980]. At first sight, Heisenberg's uncertainty relations suggest that space and time are treated in the same way in non-relativistic quantum mechanics:

$$\Delta x \,.\, \Delta p_x \geqq \frac{\hbar}{2}$$

$$\Delta y \,.\, \Delta p_y \geqq \frac{\hbar}{2}$$

$$\Delta z \,.\, \Delta p_z \geqq \frac{\hbar}{2}$$

$$\Delta t \,.\, \Delta E \geqq \frac{\hbar}{2}$$

A closer analysis reveals that this is not the case. The uncertainty relations between position and momentum are based upon the fact that the operators x and p_x (y and p_y etc.) do not commute [Jammer 1974, p.71 ff], whereas for the uncertainty relation between time and energy no such commutator relation between time t and energy H exists. Hence, the difference in interpretation and derivation of the uncertainty relations between position and momentum and between time and energy [Landau 1974, p. 158].

14 The translation is mine.

15 Cfr., for instance, F. Christensen, "The theory of space-like time", in Fraser [1978], and Christensen [1981]. Bergson has strongly objected against the tendency to treat time like space [Bergson 1944].

For the opposite question (can space be time-like?), cfr. Webb [1977].

16 For a more elaborate discussion of objective and minddependent theories of time flow, see Epilogue.

CHAPTER III

TIME ASYMMETRY

III.1 Arrow of time: time asymmetry and time flow

The arrow of time is one of the most controversial aspects of time. The meaning of the notion 'arrow of time' is rather obscure; as used in physics, this notion is somehow related to the distinction between past, present and future and the fact that the flow of time takes place in just one direction of time, from the past toward the future. Thus, the arrow of time is said to point toward the future direction of time.

Time has an arrow because nature abounds in irreversible processes. In everyday life, we come across all kinds of irreversible processes such as the mixing of coffee and cream, a stone rolling down a hill, the growth of an organism, the burning of a cigarette etc. We never observe the spontaneous reversal of these processes, viz., the spontaneous unmixing of coffee and cream, a stone rolling up a hill by itself, persons growing younger instead of older, the transformation of smoke and ashes into a cigarette. Almost all processes which we observe in everyday life are in fact irreversible; it is difficult to find examples of macroscopic processes which are perfectly reversible.

Not all of these complex irreversible processes are studied in physics; only a few kinds lend themselves to an exact, quantitative treatment. Among these are processes involving friction, such as the damped pendulum, or heat conduction ('heat only flows from the hot to the cool') and diffusion and mixing processes. The irreversible character of mixing processes is illustrated by Feynman in the following way [Feynman 1973, p. 111]:

> Suppose we have blue water, with ink, and white water, that is without ink, in a tank, with a little separation, and then we pull out the separation very delicately. The water starts separate, blue on one side and white on the other side. Wait a while. Gradually the blue mixes

> up with the white, and after a while the water is 'luke blue', I mean it is sort of fifty-fifty, the colour uniformly distributed throughout. Now if we wait and watch this for a long time, it does not by itself separate. (You could do something to get the blue separated again. You could evaporate the water and condense it somewhere else, and collect the blue dye and dissolve it in half the water, and put the thing back. But while you are doing all that you yourself would be causing irreversible phenomena somewhere else.) By itself it does not go the other way.

Indeed, no matter how long we wait, the spontaneous unmixing of both components does not occur, and whenever we repeat the experiment, the mixing of both will take place. In other words, at the phenomenological level the irreversible character of these processes appears as a law of nature.

Because of these kinds of 'one way' processes, time is said to flow in just one way: from the past toward the future. The flow of time has a fixed direction, the direction of time. The everchanging 'present moment' or 'now' steadily moves into the future: future events become present and then they become part of the past. Within the Newtonian conception of time as a straight line, this flow of time is represented by the motion of a point, the 'now' or the 'present', along this line in the direction of the future.

Without any doubt, the arrow or direction of time is one of the most intricate and obscure aspects of physical time. A cursory look at the literature on this subject is sufficient to conclude that there is much confusion about the nature of the arrow of time and it leaves the impression that the study of the arrow of time is still in a pre-paradigmatic stage. There seems to be no agreement as to what are the essential problems involved, how they are to be stated and how they could possibly be solved. A sample of the issues raised in connection with the arrow of time contains questions such as: What is, from a physical point of view, the difference between past, present and future? Why can we have traces of the past but not of the future? How can we reconcile the existence of irreversible processes with the time reversal invariant laws of classical, relativistic and quantum mechanics? How can we explain the continual growth of order and complexity in our universe? Why does time flow in just one way? Is the flow of time a physical phenomenon, i.e., is there a becoming in the physical world, or does time flow only in the realm of human consciousness? All these problems are supposed to be related in one way or another to the problem of the arrow of time.

The intricacy of the problems has caused a confusing terminology in this field, which in turn has aggravated their obscurity. At least six notions are used in close connection and sometimes as synonymous with the expression 'arrow of time', viz., 'the direction of time', 'the flow of time', 'time asymmetry', 'the anisotropy of time', 'the orientation of time' and finally 'irreversibility'. Their use is, however, by no means standardised. This makes it extremely difficult to get a hold on the problems and to compare different opinions.

In view of this, our first task will be to establish a definite terminology. In this respect, it is of fundamental importance to realise that the problem of the arrow of time contains two clusters of problems of a different nature: the problem of 'time asymmetry' and the problem of the 'flow of time'. Roughly, time asymmetry, or asymmetry in time, refers to the fact that the content of the world looks different in the two opposite orientations of time (see fig. 16a). Time asymmetry, therefore, is closely related to the occurrence of irreversible processes, for these processes introduce a real, structural difference in the content of the world relatively to both orientations of time. Consider the irreversible mixing of blue ink and water. If one looks into the future orientation of time, only the spontaneous mixing of blue ink and water will be observed, whereas in the opposite orientation of time, the same process is never observed.

The description of time asymmetry does not require the distinction between past, present and future; it is sufficient to refer to two opposite orientations of time. Thus, with regard to time asymmetry, all moments of time are equal. The flow of time, however, is based upon the distinction between past, present and future; an unique moment of time is singled out as the 'now' or the 'present'. This now flows into the direction of the future and thereby turns the near future into the near past. Fig. 16b contains a pictorial representation of the flow of time (see also Stueckelberg [1974, p. 18]).

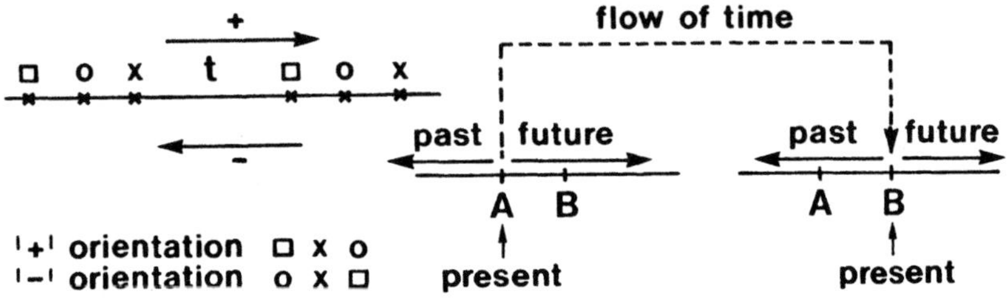

fig. 16a fig. 16b

In studies concerning the problem of the arrow of time, the distinction between time asymmetry and time flow is, in my opinion, too often neglected. Much of the confusion in the existing literature appears to be due to an ambiguous use of the notion of arrow of time.

At this place, a remark about terminology is necessary. Grünbaum [1971, p 197 ff], among others, has clearly pointed out the difference between time aymmetry and the flow of time and has also stressed the danger of confusing the two concepts; but he makes use of a different terminology. Instead of 'time asymmetry' and 'flow of time', he uses the terms 'anisotropy of time' and 'temporal becoming'. I prefer, however, to use the expression 'time asymmetry' which I adopt from Davies' [1977] book The physics of time asymmetry, because the expression 'anisotropy of time' leaves the impression that it refers to a property of time itself whereas in fact it refers to a property of the processes taking place in our universe [Davies 1977, p. 3-4, 22] (1). Moreover, unlike expressions such as the 'arrow of time' or the 'direction of time', the expression 'time asymmetry' has the advantage of avoiding any association with a flow of time (2).

The difference between time asymmetry and the flow of time can be clearly stated by means of the distinction between parameter time and coordinate time which was introduced in the previous chapter. In a spacetime manifold, events are ordered relatively to coordinate time. If this ordering is such that the contents of the spacetime manifold look different relatively to the two opposite orientations of coordinate time, then time asymmetry obtains. Thus, in a certain sense, time asymmetry pertains to coordinate time. The flow of time, however, introduces a becoming in the physical world. It can hardly be said that there is a flow of coordinate time. Relatively to coordinate time, all events and trajectories are given; there is no becoming. The introduction of a flow of time in the spacetime description of physical reality is most easily achieved by assuming that there is a flow of parameter time. This flow of parameter time corresponds to the increase of the value of parameter time along a trajectory. Due to this increase of parameter time, an observer can be said to be 'moving' along his trajectory and to have the impression of a changing universe. Therefore, there is, in the spacetime description of physical reality, a flow of parameter time and not of coordinate time.

Obviously, the two opposite orientations of time are not equivalent for irreversible processes; they take place relatively to one orientation of time only. Nevertheless, it is quite generally assumed that the fundamental laws of physics governing these processes do not distinguish between the two orientations of time. In other words, it is assumed that these laws are valid relatively to both orientations of time. Now, if this is indeed the case, i.e., if these fundamental laws are time reversal invariant, how is it then possible to explain, or at least to make plausible, the occurrence of irreversible phenomena? This problem becomes especially acute, for instance, within the context of statistical mechanics where indeed the fundamental laws are time reversal invariant; how is it possible, e.g., to account for the irreversible character of mixing processes, such as ink with water, in a purely statistical mechanical way?

All in all, time asymmetric processes are very interesting, not only physically, but also for the study of the structure of physical time. Both this and the next chapter deal exclusively with the phenomenon of time asymmetry. The main problem with regard to time asymmetry concerns the reconciliation of time asymmetry with the fundamental laws of physics. This will be the subject of Ch. IV; but first, it will be necessary to give a detailed analysis of the notion of time asymmetry itself. That will be done in the next section, where I shall propose a probabilistic definition of time asymmetry, which will be supplemented, in section III.3, by a discussion of the time reversal operator T*. There I shall also discuss what it means to reverse a process in time, and when a law of physics is time reversal invariant. A clear insight into these two issues is of primary importance for an understanding of the physical status of time asymmetry. Finally, I shall investigate in section III.4 the relation between time asymmetry and temporal orientability of a spacetime manifold; this is necessary because the definition of time asymmetry presupposes a distinction between two opposite time orientations.

III.2 Time asymmetry

III.2.1 A probabilistic conception of time asymmetry.

In this section I shall proceed in the following way. First, I shall present two definitions of time asymmetry as found in the literature. A close examination of these shows that both have serious shortcomings; one of the main objections against both of them is that they put too much emphasis upon the role of irreversible processes. A more general definition of time asymmetry will then be proposed. In the next section I shall try to show that, although this definition is itself not without problems, it can nevertheless serve as a useful point of departure for further inquiry.

Davies [1977, p. 3-4] describes time asymmetry as follows:

> Time asymmetry is here taken to mean the basic fact of nature that the contents of the world possess a structural distinction between past and future facing orientations.

He strongly underlines the fact that time asymmetry is not a property of time itself but of the contents of the world. Therefore, he explicitly rejects the expression 'asymmetry of time' [op. cit. p. 22].

Davies speaks of a structural distinction between the two opposite time orientations and gives the following elucidation of what this means [op. cit. p. 31]:

> The two directions of time in the following sense - toward the past and toward the future - are known from common experience to be fundamentally distinguished physically...For example, we remember the past. Moreover, this asymmetry with respect to the two time orientations is also readily recognized in laboratory physics. Indeed, practically all phenomena of nature appear to be asymmetric in time.

This description of time asymmetry, however, does not help us very much to understand the precise nature of the structural distinction between past and future orientation of time. Davies' example, that we can remember the past, is at the very least problematic since it refers to the notion of the past which, as we have seen in the introduction, is related to the flow of time, not to time asymmetry. Neither does his remark that "practically all phenomena of nature appear to be asymmetric in time" bring us any further, since it is not clear what it means for phenomena to be asymmetric in time. Apparently this asymmetry is related to irreversibility, for Davies [op. cit. p. 34] claims that "thermodynamic macroscopic irreversibility does imply an asymmetry in time." Thus, irreversible macroscopic phenomena are asymmetric in time.

The qualification 'macroscopic' in 'thermodynamic macroscopic irreversibility' is important, for Davies distinguishes between macroscopic and microscopic irreversibility; whereas macroscopic irreversibility does imply time asymmetry, this is not necessarily the case for microscopic irreversibility [ibidem]. This is due to the fact that time asymmetry, according to Davies, is a valid concept only within a macroscopic description of the physical world, because it refers to a "collective property of physical systems in space-time" [op. cit. p. 3].

I shall not follow Davies in his restriction of the use of the notion 'time asymmetry' to the macroscopic description of the physical world. It is possible to define time asymmetry, as will be seen shortly, so that it can be applied coherently to microscopic systems as well. Consequently, there is no need to introduce, as Davies does, a distinction between macroscopic and microscopic irreversibility. Moreover, irreversibility can be defined so that the size of the systems involved is irrelevant (see sect. III.3.2).

Irreversibility also plays a crucial role in Grünbaum's description of time asymmetry (for which he uses the term 'anisotropy of time') [1971, p. 200-201]:

> the existence of irreversible processes structurally distinguishes the two opposite time senses as follows: there are certain kinds of sequences of states of systems specified in the order of decreasing time coordinates such that the same kinds of sequences do not likewise obtain in the order of increasing time coordinates. Accordingly, if there are irreversible kinds of processes, then time is anisotropic [asymmetric].

There can be no doubt about what constitutes, according to Grünbaum, a structural distinction between the two opposite time orientations. Again, irreversibility is the key concept; time asymmetry is based upon the existence of irreversible processes.

These irreversible processes can be of two different types. Following Mehlberg, Grünbaum [1973, p. 210-211] distinguishes between weak or de facto irreversible processes and strong or nomological irreversible ones. De facto (weak) irreversibility obtains when the reversal of a process is never or almost never observed, but is not forbidden by the laws of physics, whereas in the case of nomological (strong) irreversibility the reversed process is forbidden by those laws (3).

According to Grünbaum, time asymmetry obtains in case nomological or de facto irreversible processes occur; in this manner, he creates the impression that, in one way or another, irreversibility is a necessary condition for time aymmetry. Thus, the emphasis is put upon the existence of irreversible processes; this view is but one step away from the identification of time asymmetry with irreversibility. It is precisely this standard coupling of time asymmetry with irreversibility that I shall try to avoid by defining time asymmetry without any explicit reference to irreversibility. The feasibility and usefulness of such definition appears from the following considerations.

Clearly, the existence of irreversible processes, whether de facto or nomological, invalidates the equivalence of the two orientations of time. But the interesting question is whether the existence of irreversible processes is a necessary or a sufficient condition for the occurrence of time asymmetry. It appears to be only a sufficient condition, for perfectly reversible processes too can furnish a basis for time asymmetry. Suppose that instances of a kind of process A occur relative to both orientations of time; in other words, that A is reversible. Consider the case in which the probability for the occurrence of instances of A relative to the past orientation of time differs from the probability for the occurrence of instances of A relative to the future orientation. Then, there is a statistical difference in the contents of the world relative to past and future orientation of time, and it is possible to distinguish unambiguously between the two orientations of time on the basis of the probabilities for the occurrence of processes of the kind A; past and future orientation of time are not equivalent. Thus, time asymmetry obtains with regard to processes A, although A is reversible and de facto reversed. Grünbaum considers only a special case of time asymmetry, in which one of the probabilities is zero or practically zero, that is, the case in which the proceses involved are nomologically or de facto irreversible.

Hence, the following definition of time asymmetry appears to be more appropriate:

> Time asymmetry is the basic fact of nature that the content of the world shows a structural difference between past and future facing orientations of time in the sense that there exist certain kinds of processes A, B, C..., such that the probability for the occurrence of instances of A, B, C.., relatively to the future orientation of time, denoted by $P^{+}(A)$, etc., differs from the probability for the occurrence of the same kinds of processes relatively to the past orientation of time, denoted by $P^{-}(A)$ etc.; that is
>
> $$P^{+}(A) \neq P^{-}(A), \text{ etc.}$$

This definition has the advantage of making a clear separation between time asymmetry and irreversibility; the existence of time asymmetry is not exclusively linked to the existence of irreversible processes (4).

It is easy to see that Grünbaum's description of time asymmetry in terms of irreversible processes is just a special case of our definition. In the limit case, where $P^+(A) \approx 0$ and $P^-(A) > 0$ (that is, there is a real chance that instances of A occur relatively to the past orientation of time, but not relatively to the future orientation), we recover the case of de facto irreversible processes. The corresponding type of time asymmetry will be called 'de facto time asymmetry'. In the case where $P^+(A) > 0$ and $P^-(A) = 0$ (the occurrence of instances of A relatively to the past orientation is forbidden by the laws of physics), nomological irreversible processes are involved; the corresponding type of time asymmetry will be called 'nomological time asymmetry'.

III.2.2 Discussion of probabilistic time asymmetry.

The apparent innocence of our definition of time asymmetry cannot conceal some real problems which require closer examination. First, and foremost, there is the problem of defining the probabilities $P^+(A)$, $P^-(A)$ etc. Secondly, an important time asymmetric process, viz., the expansion of the universe, is not covered by this definition. Finally, it will be necessary to reformulate this definition so that it contains only probabilities relatively to the future facing orientation of time. In other words, we must redefine the probabilities $P^-(A)$, $P^-(B)$ etc. into probabilities relatively to the future orientation: $P^-(A) \longrightarrow P^+(?)$. A closer examination of these topics, especially the first, shows that the above definition of time asymmetry is more complicated than might be expected at first sight.

Probabilities $P^+(A)$ etc. as relative frequencies.

In our definition, time asymmetry is a probabilistic concept; therefore, it agrees very much with Reichenbach's conception of time asymmetry. According to Reichenbach, time asymmetry is "a matter of probability and statistics" and "the statistical nature of time direction [time asymmetry] appears to be the ultimate outcome of all inquiries into the nature of time" [1971,

p. 134]. Because time asymmetry is a matter of probability, it requires reference to an ensemble of systems in order to secure that the relevant probabilities are well defined. Here we have to face the following question: Under what conditions can a collection of physical systems be considered as an ensemble so that we may be confident about the probabilities determined on the basis of this collection of physical systems (5)?

This is an extremely complex problem. In his Direction of time, Reichenbach [1971] studies in detail the statistical behaviour of ensembles of 'branch systems' and shows that, under certain conditions, such ensembles behave asymmetrically in time (see sect. IV.2.4). But he says but little about the question regarding the conditions under which a collection of branch systems can be considered as an ensemble.

Roughly, the problem concerning the determination of the probabilities can be stated in the following way. In reality, we never measure directly probabilities, but only relative frequencies. The question now is: under what conditions can inferences about probabilities be made on the basis of observed relative frequencies. In the case of time asymmetry, the problem is the following. Consider a finite region of spacetime $\Delta\Omega = \Delta S x \Delta T$, and suppose that a large number N of processes of the kind A take place in $\Delta\Omega$. Note that N is the sum of the number of instances of A relative to the future orientation of time, n^+, and the number of instances of A relatively to the past orientation, n^-. The relative frequencies with regard to past and future orientation are given by:

$$F^-(A) = n^-/N \qquad F^+(A) = n^+/N.$$

Now the problem can be stated in a simple way: under what conditions is $F^-(A)$ etc. an adequate measure of P^- (A) etc.? This is one of the central epistemological problems of probability theory for which there is as yet no straightforward solution. As long as these conditions stay in the dark, the determination of the probabilities $P^+(A)$ etc., and consequently also the probabilistic conception of time asymmetry, remains problematic.

These difficulties with regard to the determination of the relevant probabilities can be avoided by simply restricting the use of the notion of time asymmetry to finite regions of spacetime. In that case, the probabilities can in principle be given in terms of the relative frequencies, as was indicated above. Time asymmetry is then considered to be a regional phenomenon. This, however, is in accordance with what is to be expected if time asymmetry is of a de facto nature, as it is almost generally assumed. In that case, the time asymmetric behaviour of systems is not the result of the laws of physics but of the prevailing boundary conditions in that part of the universe where these time asymmetric processes occur (see sect. III.4 and Ch. IV). These boundary conditions do not necessarily obtain everywhere on spacetime.

Following this line of thought, the existence of time asymmetry is restricted

a) to particular kinds of processes A, B, C,..

b) to (finite) regions of space,

c) to (finite) spans of time.

This restriction of time asymmetry to (finite) regions of spacetime, however, does not solve all our problems. It works quite well for those processes whose reversal is (almost) never observed, like the mixing of ink and water. Their time asymmetric behaviour is evident from direct observations; for any 'known' spacetime region, the mixing of both is always observed relatively to the future orientation, and the spontaneous unmixing only relatively to the past orientation. The inductive inference is made to the lawlike irreversible character of these mixing processes because we never come across counterinstances.

But suppose that a comparison of the relative frequencies $F^{+}(A)$ and $F^{-}(A)$ for a finite region of spacetime $\Delta\Omega$ shows that they are just slightly different. In that case, the time asymetric behaviour of processes of the kind A is much more problematic. Then we must make sure that the relative frequencies are not dependent upon the choice of $\Delta\Omega$; if they are, then a different choice for $\Delta\Omega$ might lead to just the opposite time asymmetric behaviour of processes of the kind A.

The time asymmetric phenomena observed in daily life mostly concern processes whose reversal is never observed. It would be wrong, however, to conclude that time asymmetries with regard to processes for which the reversal does actually take place, and for which $F^{+}(A)$

and $F^-(A)$ are only slightly different, are not physically relevant. Such processes might have played an important role in the evolution of the universe, especially in the early stages.

Consider the stage in the evolution of the universe at which there still was thermal equilibrium between radiation, and electrons and positrons. Here, the temperature of the universe is still sufficiently high (> 6.10^9 ^oK) for electron-positron creation from radiation to occur:

A: $2\gamma \longrightarrow e^+ + e^-$ (electron-positron creation).

In the situation of thermal equilibrium, the rate of creation of electron-positron pairs from radiation equals the rate of annihilation of those pairs; this annihilation is considered to be the reversal of process A and is denoted by A^r:

A^r: $e^+ + e^- \longrightarrow 2\gamma$ (electron-positron annihilation).

In other words, the rate of creation of electron-positron pairs is the same for both orientations of time:

$$P^+(A) = P^-(A).$$

Thus, during this stage of the evolution of the universe, there is no time asymmetry with regard to processes of type A.

Due to the expansion of the universe, its temperature drops until at a certain moment it reaches a value which is about equal to the treshold temperature for the creation of electron-positron pairs from radiation. Above this value, gamma-rays do contain enough energy for the creation of electron-positron pairs, and the probability for the occurrence of this process is high enough to maintain equilibrium. Below this value, the probability for the creation of electron-positron pairs rapidly decreases. Thus, in the process of cooling down, the universe reaches a point at which the equilibrium between radiation and electrons and positrons can no longer be maintained. From that moment on,

$$P^+(A) < P^-(A).$$

Because of the changing boundary conditions in our universe, the probabilities become unequal and from that moment on, time asymmetry obtains relatively to processes of the kind A. At the beginning, there is only a slight difference between $P^+(A)$ and $P^-(A)$, but as the temperature drops further, the discrepancy between them increases, until the probability for the creation of electron-positron pairs becomes almost zero; from that moment on, only annihilation takes place until electrons and positrons "disappear as major constituents of the universe" [Weinberg 1978, p. 107].

This example clearly illustrates that i) time asymmetry with regard to perfectly reversible processes is possible, ii) time asymmetry can be caused by changing boundary conditions in the universe, and iii) the occurrence of time asymmetry can be limited to a finite time interval during which the relevant probabilities might be only slightly different.

Cosmological time asymmetry.

It is evident that the given definition of time asymmetry cannot be applied to processes such as the expansion of the universe; in those cases a probabilistic approach to the problem of time asymmetry is not appropriate. Nevertheless, the expansion of the universe is clearly a time asymmetric process; as we shall see in chapter IV, it is sometimes even regarded to be the most fundamental time asymmetric process to which all other kinds of time asymmetry are to be reduced. For example, the asymmetry with regard to the creation and annihilation of electron-positron pairs, discussed above, is caused by the expansion of the universe. The starting point for the probabilistic conception of time asymmetry was not, however, this overall time asymmetric process, but the time asymmetric behaviour of physical systems as observed in daily life. This time asymmetric behaviour can best be characterised in a probabilistic way in spite of the fact that, may be, ultimately this time asymmetric behaviour is due to the time asymmetric expansion of the universe.

Reformulation of $P^-(A)$ into $P^+(?)$.

The definition of time asymmetry contains probabilities relative to the past orientation of time. We cannot, however, verify directly whether a process A takes place relative to the past orientation of time; we cannot go 'back in time' in order to see whether processes of the kind A take place relative to the past orientation of time. What we can observe is whether both A and the reversal of A take place with regard to the future orientation of time. This means that the definition must be restated so that it contains only probabilities relative to the future orientation of time. This can easily be achieved, since the probability for the occurrence of instances of A relative to the past orientation of time is by definition equal to the probability for the occurrence of instances of the reversal of A, A^r, relative to the future orientation of time. Therefore the inequality

$$P^-(A) \neq P^+(A),$$

can be rewritten as

$$P^+(A^r) \neq P^+(A).$$

Thus, time asymmetry obtains with regard to processes of the kind A if the probabilities for the occurrence of instances of A and A^r are different, provided the probabilities are taken relatively to the same orientation of time.

While it is intuitively quite clear what is meant by the 'temporally reversed process of A' in cases such as mixing processes - the reversal of the spontaneous mixing of blue ink with water would be the spontaneous unmixing of both - the precise physical definition of the reversal of a process is far more difficult. To this intriguing problem I shall return in the next section, since it is intimately connected with the nature of the time reversal operator T*.

In spite of the foregoing difficulties, the probabilistic conception of time asymmetry can serve as a promising starting point for further analysis. The main advantage of this conception of time asymmetry is, as I have already remarked, that it disconnects time asymmetry from irreversibility, contrary to the standard conceptions of time asymmetry which are based upon irreversibility. It is not necessary to identify time asymmetry with irreversibility; time asymmetry can obtain even if all processes are reversible. Therefore this conception of time asymmetry is clearly compatible with the alleged time reversal invariance of the fundamental laws of physics, whereas, if time asymmetry is based exclusively on irreversible processes, there is at least a tension between the existence of time asymmetry in our universe and the time reversal invariance of physical laws.

The probabilistic conception of time asymmetry displays yet other interesting features. I have already pointed out that it is compatible with the idea that time asymmetry is a regional phenomenon. Equally important are the following two points: a) time asymmetry as defined above is not necessarily a macroscopic phenomenon, and b) time asymmetry is in principle not a property of individual processes, but of ensembles of processes.

ad a) The first point concerns the relation between time asymmetry and the use of a macroscopic or a microscopic level of description of physical reality. In the literature it is quite often remarked that time asymmetry is only relevant for a macroscopic description of the world. According to this point of view, only macroscopic systems can behave asymmetrically in time. Time asymmetry is considered to be meaningless in a microscopic description because all microscopic processes are reversible: a clear example of the standard coupling of time asymmetry with irreversibility! Consider again the mixing of blue ink with water. The collisions between the individual ink and water molecules are all perfectly reversible. Therefore it is claimed that the time asymmetric behaviour can only be a property of the collective behaviour of the molecules, which shows itself at the macroscopic level. A typical proponent of this view is Davies [1977, p. 4]:

> time asymmetry is only a type of description, relevant to the macroscopic worldview of the physicist. But nothing yet discovered in nature requires individual atoms to experience time asymmetry, the very essense of which is the collective quality of complex systems.

Starting from the probabilistic conception of time asymmetry, there is no reason why time asymmetry should not or could not occur with regard to microscopic processes. We have already seen that it can be applied meaningfully to the microscopic processes of creation and annihilation of electron-positron pairs. Consider also the following radioactive decay of ^{238}Uranium atoms:

A: $^{238}U \longrightarrow {}^{234}Th + \alpha$-particle.

Instances of A frequently take place in our region of the universe, but instances of A^r almost never. Therefore, $F^+(A) \gg F^+(A^r)$, and thus, time asymmetry obtains in our region of the universe for this kind of microscopic processes. In spite of the fact that this process is usually regarded to be reversible, since the laws governing this type of radioactive decay are supposed to be time reversal invariant, it nevertheless shows time asymmetric behaviour which is certainly not the result of the collective behaviour of many interacting molecules.

Thus, it is not necessary to restrict the use of the notion of time asymmetry to the macroscopic description of the world. On the other hand, I do agree with Davies' remark that time asymmetry is not a quality of individual atoms or, more generally, of individual systems; this brings me to the second point.

ad b) As defined here, time asymmetry is not a property of individual systems or processes. Time asymmetry is essentially a probabilistic concept; this means that an individual process of the kind A is not asymmetric in time, but that time asymmetry obtains in the ideal case with regard to ensembles of processes A. Of course, in the limiting case where the probability for the reversed process is zero or approximately zero, it will be possible to distinguish the past from the future orientation of time by a single instance of A. For example, when we are shown two pictures, one of blue ink and water in an unmixed state and another of the same substances in a mixed state, we can infer with absolute certainty the temporal order in which both pictures were taken, provided the system 'blue ink and water' was closed between the two moments, at which these pictures were taken. In those limiting cases, the time asymmetric behaviour becomes a property of the individual systems.

The same line of argument remains valid if we start from two pictures showing, respectively, an 'ordered' and an 'unordered' deck of playing cards of which one was produced from the other by shuffling. The chance that shuffling an unordered deck of cards will produce an ordered one is extremely small, and therefore we infer that the picture of the ordered deck of cards was taken before the picture of the unordered one. But with the help of this example, it is very easy to show that the number of cards in a deck determines to a large degree the possibility of ascribing a time asymmetric character to the process of shuffling under the given conditions. Indeed, a decrease in the number of cards in the deck will increase the probability for the spontaneous occurrence of the reversed process, from an unordered to an ordered deck of cards. In case the deck contains only a few cards, there is a fair chance that the picture of the unordered deck was taken before the ordered one, because shuffling that small number of cards could very well produce an ordered deck of cards from an unordered deck. This demonstrates clearly the probabilistic nature of the time asymmetric shuffling process.

Still one problem with regard to our definition of time asymmetry remains to be discussed, namely the definition of the reversed process A^r of A; this will be done in the next section. There we will see that, from a physical point of view, the definition of A^r is not at all trivial.

III.3 (Ir)reversibility and the time reversal operator T*

III.3.1 Time reversal.

Although our conception of time asymmetry is not based explicitly on irreversible processes, it goes without saying that irreversible processes are of primary importance with regard to time asymmetry as observed in daily life. My intention so far has not been to claim that irreversibility is irrelevant for time asymmetry; instead, I have tried to clarify the relation between the two concepts in order to avoid the widespread identification of time asymmetry with irreversibility.

In everyday life, it is quite clear what is meant by irreversible processes; processes are called irreversible if the reversed process cannot occur. Usually, we tacitly assume that we all know what constitutes the reversal of a process. In case of mixing processes, it is the spontaneous unmixing of the components; in case of the motion of a damped pendulum, it is a pendulum that starts swinging spontaneously with an ever greater amplitude, etc. In this section, however, it will be seen that, from a physical point of view, the definition of the reversed process is rather complicated, and that the time reversal operator T* plays a fundamental role in this issue.

A turning point in the study of irreversible processes in physics is the work of Carnot, Clausius and Thomson on heat phenomena, which dates from the 19th century. They formulated the second law of phenomenological thermodynamics which states that the entropy of a closed system increases or remains equal, but never decreases. This entropy concept permits a thermodynamical classification of processes into two types: reversible and irreversible. Processes whose entropy does not change are called reversible, while those whose entropy increases are called irreversible, provided the corresponding systems are closed. This has become the standard approach in physics to the problem of classifying processes as reversible and irreversible, and that is the reason why the concept of entropy plays such a dominant role in discussions about the nature and origin of time asymmetry (see Ch. IV).

Here I shall follow a different course for two reasons. First, this classification of processes on the basis of the concept of entropy is rather foreign to our experience (we do not directly observe entropy in nature). But secondly, and more importantly, this classification procedure is not adequate. In the next chapter, we will encounter irreversible processes in which the role of entropy is rather controversial.

A precise definition of irreversibility can be given without reference to the notion of entropy. Given a closed system Σ; let this system go through an evolution and let this evolution be described physically by a parametrised curve in some phase space (see Ch. II). In other words, system Σ goes throuh a process A which can be decomposed into a sequence of states in some phase space, parametrised by time t:

$$A: \quad A(t_i)\ldots\ldots A(t)\ldots\ldots A(t_f), \qquad t_i < t < t_f .$$

$A(t_i)$ and $A(t_f)$ are, respectively, the initial and final state of the system. Now let us consider the process A from a time reversed point of view. Clearly, the temporal order of the states will be reversed. But this is not the only thing that will change; the states themselves will also look different if time is reversed. For instance, if the state of the system Σ is given in terms of position and momentum of its component particles, then, from a time reversed point of view, the positions of the particles will be the same, but their momentum will change sign because, intuitively, time reversal corresponds to motion reversal.

Generally, the state $A(t)$ will be different from its time reversed counterpart, $A^r(t)$. $A(t)$ represents the states of the system Σ as observed moving in the future orientation of time, and $A^r(t)$ as observed going into the past orientation. The relation between $A(t)$ and $A^r(t)$ is given by the time reversal operator T^*; $A^r(t)$ is the result of the action of T^* on $A(t)$:

$$A^r(t) = T^*A(t).$$

Thus if system Σ, relatively to the future orientation of time, goes through a process A, then from the time reversed point of view, system Σ goes through the process A^r given by the following sequence of states:

$$A^r: \quad A^r(t_f)\ldots.A^r(t)\ldots.A^r(t_i).$$

The temporal reverse of a process is therefore not just the reversed temporal order of the states of the original process, but it is given by the temporally reversed states in reversed order.

III.3.2 (Ir)reversibility and time reversal invariant laws.

The definition of reversibility follows immediately: a process A is called reversible if both A and A^r are allowed by the laws of physics to occur relatively to the future (or the past) orientation of time. If not, the process is irreversible.

The assumption that system Σ is closed during the process A also applies to the reversed process A^r. More generally, if certain constraints $C^1,\ldots,C^n$ apply to the system Σ when it goes through the process A, the same constraints apply to the time reversed process, but as seen from the time reversed point of view, that is $T^*(C^1),\ldots,T^*(C^n)$.

Note that the reversibility of a process A does not by itself imply that in reality the process will be reversed. The occurrence of the appropriate initial conditions is of primary importance (see Ch. IV). If, relatively to the future orientation of time, the initial state corresponding to $A^r(t_f)$ de facto never or seldom occurs, but the one corresponding to $A(t_i)$ does, then the reversal of A will never or seldom take place. In accordance with what was said at p. 110, A is in that case called 'de facto irreversible'. Yet, this is rather misleading since, according to the above definition, these processes are reversible in spite of the fact that the reversed process does not take place in reality (6). Landsberg [1970, p. 1122] characterises de facto (weakly) irreversible processes as 'weakly T-invariant', that is, their time reversal, although improbable, does "not violate the laws of the most elementary processes in terms of which it is fully understood".

Whether or not a process A is reversible, depends upon the character of the laws governing that kind of process. It will be clear by now that, if the same laws are valid relatively to both orientations of time, the processes governed by these laws will be reversible. In other words, time reversal invariance of physical laws implies reversibility. It is not necessary to actually reverse the course of time or to let time, so to say, flow backward in order to test the time reversal invariance of physical laws. Sklar describes the following procedure for determining whether the laws governing a given system are time reversal invariant [1977, p. 367]:

> We start off with a system in state S, allowing it to evolve, after time Dt to state S_1. At the same time we start off with another system, exactly like the first, except that its initial state is the 'time-reversed state' of the final state of the original system. Call this new state $T(S_1)$. If the laws of nature are time-reversal invariant, then at the end of interval Dt we will find the second system in the state T(S), the time-reversed version of the original state of the first system.

It is, however, crucial to choose the proper time reversed state and to insure that all the constraints are also taken from the time reversed point of view.

III.3.3 The time reversal operator T*.

In order to determine the correct time reversed state $A^r(t)$ of $A(t)$, the action of T* on A(t) must be known; this requires that the action of T* on all physical quantities necessary for completely specifying the state of system Σ is known. In that case, the time reversed process of A is also uniquely determined. Moreover, if the behaviour of all the quantities appearing in the laws governing the system Σ under T* is known, then it can also easily be determined whether or not process A is reversible.

The time reversal operator T* transforms t into -t; it does not affect the position $\bar{x}$ - in the following, 'barred' quantities denote vectors - of a particle in space:

$$T^*((\bar{x},t)) \longrightarrow (\bar{x},-t).$$

Therefore, the velocity $\bar{v}$ of a particle also changes sign under time reversal:

$$T^*(\bar{v}) \longrightarrow -\bar{v}.$$

Thus, time reversal corresponds to motion reversal. Accordingly, the time reversal operator T* is usually defined in physics as the operator that changes the sign of momentum $\bar{p}$, and spin $\bar{s}$ [Sakurai, 1973, p. 173]:

$$T^*(\bar{p}) \longrightarrow -\bar{p},$$
$$T^*(\bar{s}) \longrightarrow -\bar{s}.$$

Thus, if the state of a particle is given by position, momentum and spin, $A=(\bar{x},\bar{p},\bar{s})$, then the action of T* on A results in a change of sign of the momentum and the spin.

Knowledge of the action of T* on position, momentum and spin does not, however, suffice to determine whether or not a process is reversible; beside that, the effect of T* on every physical quantity appearing in the laws governing that process should be known. Usually, the action of T* on all these quantities is not explicitly given. The action of T*, however, should be defined coherently for all physical quantities involved. The resulting problem is best illustrated by the following two examples, 1) the second law of Newton which has become the standard example of a time reversal invariant law, and 2) the second law of phenomenological thermodynamics which is the standard example of a time reversal non invariant law.

1) Newton's second law.

Within classical mechanics, the acceleration $\bar{a}$ of a particle with mass m under a force $\bar{F}$ is given by Newton's second law:

$$\bar{F} = m.\bar{a} = m.d^2\bar{x}(t)/dt^2.$$

The state of a particle at time t_0 is completely determined by its place and velocity, $A(t_0)=(\bar{x},\bar{v})$; thus $A^r(t_0)$ is given by $(\bar{x},-\bar{v})$.

The action of T* on this second law is determined as follows. In the first place, a formal rule must be laid down:

$$T^*(x.y) = T^*(x).T^*(y),$$

which says that T* is a morphism with regard to multiplication (and addition). Now, applying T* to the second law yields:

$$T^*(\bar{F}) = T^*(m).T^*(\bar{a}).$$

Since the acceleration $\bar{a}$ is the second derivative with regard to t, it does not change sign under T*. What about the action of T* on the mass m? The invariance of m under T* follows from the definition of the action of T* on the momentum $\bar{p}$: $T^*(\bar{p}) = -\bar{p}$; this implies that $T^*(m) = m$. Note, however, that the invariance of m under T* is based upon a convention which is hidden in the assumption that the momentum $\bar{p} = m.\bar{v}$ does change sign under time reversal. This assumption is not inherent to the conception of time reversal as motion reversal; there is no intrinsic reason why m could not change sign under T*. Assuming that m does change, and thus that $\bar{p}$ does not change sign under T*, the action of T* on all other physical quantities involving m must be reconsidered in order to be coherent; in that case, e.g., the (kinetic) energy of a particle also has to change sign under T*.

Given the standard convention that m does not change sign under T*, the time reversal invariance of the second law of Newton depends upon the action of T* on

the force $\bar{F}$. I shall briefly consider two cases: i) the gravitational force, and ii) the Lorentz force.
i) In case $\bar{F}$ stands for the gravitational force,

$$\bar{F} = (Gm_1m_2)/r^2$$

application of T* shows that it does not matter whether mass changes sign under T* but also that the gravitational force is invariant under time reversal provided the gravitational constant G does not change sign under T* (if G were to change sign under time reversal the attractive gravitational force would become a repulsive force). Thus, given the conventions that T*(m) = m and T*(G) = G, Newton's second second law is time reversal invariant in case $\bar{F}$ refers to the gravitational force.
ii) With regard to the Lorentz force, acting on a particle with electric charge e moving with velocity $\bar{v}$ in a magnetic field B,

$$\bar{F} = e.\bar{v} \times \bar{B}$$

the invariance under T* is usually made plausible by arguing that, beside $\bar{v}$, $\bar{B}$ also changes sign under T*, since "the magnetic field itself is produced by moving charges, and under time reversal of these charges the field changes sign" [Davies 1977, p. 26]. This is true, however, provided it is tacitly assumed that the electric charge e does not change sign under T*, i.e, T*(e) = e (7). Then, indeed, the Lorentz force, and with it Newton's second law, is time reversal invariant:

$$T^*(\bar{F}) = e.(-\bar{v})x(-\bar{B}) = \bar{F}.$$

But as in the case of the mass, the question arises why the assumption should be made that electric charge does not change sign under T*.
- Intermezzo: a geometrical interpretation of the CPT* theorem.
It is not at all obvious that e does not change sign under T*, especially in view of Feynman's and Stueckelberg's interpretation of a positron as an electron 'going back in time' (8), which suggests that e does change sign under T*. With regard to this problem, the distinction between parameter time and coordinate time is relevant, since the action of time reversal on e can be defined differently for parameter time (or proper

time) and for coordinate time. Let us assume that coordinate time reversal T_c^* does not change the sign of e:

$$T_c^*(e) = e$$

but that parameter time reversal T_p^* does change the sign of e, as suggested by the Feynman-Stueckelberg interpretation of a positron:

$$T_p^*(e) = -e.$$

This distinction between T_c^* and T_p^* has interesting consequences for the interpretation of the CPT* theorem. If T_p^* does change the sign of electric charge, then it has the same effect on e as the charge conjugation operator C. But in the CPT* theorem, C has a more general interpretation in the sense that it corresponds to particle-antiparticle conjugation.

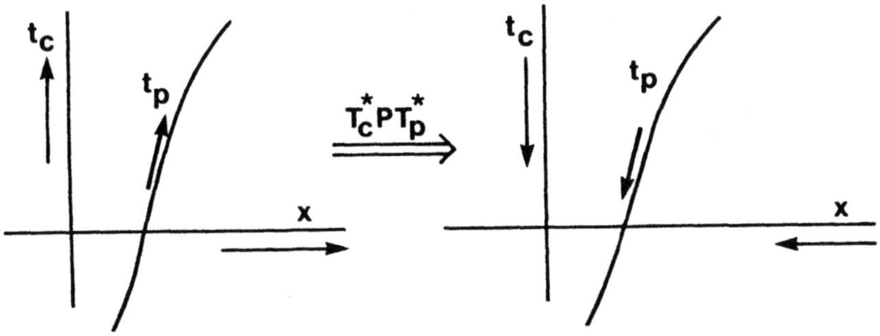

fig. 17

Scheurer [1970] has suggested to interpret 'parameter time reversal' in the same way as 'corresponding to particle-antiparticle conjugation'. With this interpretation of charge conjugation as parameter time reversal, the CPT* of physical laws receives a purely geometrical meaning. This becomes immediately clear if the time reversal operatot T* in the CPT* theorem is seen as referring to coordinate time reversal T_c^*; assuming the equivalence of C with T_p^*, we arrive at the $T_p^* P T_c^*$ invariance of physical laws: a completely reversed description of, e.g., the motion of a particle, requires reversal not only of the orientations of the space coordinates (P) and time coordinate (T_c^*), but also of the direction of the increase of parameter time along the trajectory of the particle (T^*) (see fig. 17).

2) The second law of phenomenological thermodynamics.
The standard example of a time reversal non invariant law is the second law of phenomenological thermodynamics, usually given as

$$dS(t)/dt \geq 0,$$

where $S(t)$ is the entropy as a function of time. But whether or not this law is indeed time reversal non invariant depends upon the action of T^* on S. There are two options here:

a) $T^*(S) = S$. This is the standard convention which leads to a time reversal non invariant second law:

$$T^*(dS/dt) = -dS/dt \geq 0; \quad \text{---->} \quad dS/dt \leq 0.$$

Because

$$dS = dQ/T,$$

where dQ is the heat supplied to the system, and T is the absolute temperature, T does not change sign under time reversal, $T^*(T) = T$, provided that heat, as a form of energy, does not change sign under T^*. Similarly, the action of T^* on all thermodynamic quantities, like heat capacity and friction coefficient, has to be defined.

b) $T^*(S) = -S$. If this convention is adopted, then the second law of phenomenological thermodynamics turns out to be time reversal invariant:

$$T^*(dS/dt) = dS/dt \geq 0.$$

A consequence of this definition of the action of T^* on S is that absolute temperature has to change sign under T^* as well. Although it is not standard to assume that $T^*(S) = -S$, Stueckelberg [1974] has shown that this assumption is possible, provided the action of T^* on all other thermodynamical quantities is coherently defined (9).

Thus, we may conclude that whether or not the second law of phenomenological thermodynamics is time reversal invariant, depends upon which convention is adopted for the action of T^* on the thermodynamic entropy S.

These two examples suffice to show that it is not always self-evident how the operator T* acts on physical quantities. In principle, a coherent definition of the action of T* on physical quantities can be achieved in several ways. As in these two examples, the time reversal operator T* must be defined explicitly for every domain of physics in order to determine whether the laws in those domains are time reversal invariant (10).

III.3.4 (Ir)reversibility and levels of description.

It is important to realise that the evolution of a system can be described in different ways depending upon the characterisation of the state of the system. This implies that the time reversal of a given process A is not uniquely defined, which again has important consequences for the reversibility or irreversibility of a process.

Consider the situation in which the state of a gas is given in terms of pressure P, absolute temperature T and volume V:

$$A = (P,T,V).$$

Following the standard convention for the action of T* on P, T and V, we obtain:

$$T^*(A) = A^r = (P,T,V) = A.$$

Thus the time reversed state of the system is identical to the original state. But the same system can be described by specifying the places $\bar{x}_i$ and the momenta $\bar{p}_i$ of the molecules of the gas:

$$A' = (\bar{x}_i,\bar{p}_i),$$

with i = 1,....,N, N: the number of molecules in the gas. In this description, the time reversed state T*(A') is not identical with the original state:

$$T^*(A') = A'^r = (\bar{x}_i,-\bar{p}_i) \neq A.$$

This shows that it is not always meaningful to speak about the time reversed state of a system Σ as such. The nature of the time reversed state depends upon the (dynamical) quantities by which the state of system Σ is characterised. The same applies to the reversal of a process; the nature of the temporal reversal of the evolution of a system depends upon the mode of description of the evolution of the system. For example, if the evolution of a gas is described in terms of P, T and V,

$$A: \quad (P_0, T_0, V_0) \longrightarrow (P_1, T_1, V_1),$$

then the time reversed process is given by:

$$A^r: \quad (P_1, T_1, V_1) \longrightarrow (P_0, T_0, V_0).$$

But if, on the other hand, the same evolution of the gas is described in terms of places and momenta of the individual molecules, then

$$A': \quad S_0 = (\bar{x}_i^0, \bar{p}_i^0) \longrightarrow S_1 = (\bar{x}_i^1, \bar{p}_i^1),$$

and

$$A'^r: \quad S_1^r = (\bar{x}_i^1, -\bar{p}_i^1) \longrightarrow S_0^r = (\bar{x}_i^0, -\bar{p}_i^0).$$

This has far-reaching consequences for the problem of (ir)reversibility. The evolution of a system may be irreversible in one description and reversible in another. Consider a bar with an initial temperature distribution as given in fig. 18a. According to the macroscopic law of heat conduction,

$$\delta T/\delta t = \kappa\, \delta^2 T/\delta x^2,$$

with κ the coefficient of heat conduction ($\kappa > 0$), the system will have, after a period Δt, a temperature distribution as given in fig. 18b. According to the same law, the time reversal of this process, i.e., the transition from the state of fig. 18b to the one of fig. 18a, is forbidden. In this macroscopic description, the process is irreversible, due of course to the time reversal non invariance of the macroscopic law of heat conduction.

In a microscopic description of the same process in terms of places and momenta of the molecules constituting the bar, the situation becomes quite different. Assume that in this microscopic description the initial state is given by S_0 and the final state by S_1. According to the laws of classical mechanics, the system evolves deterministically in time Δt from state S_0 to state S_1. But the same applies to the reversed process which is given by the transition from state S_1^r to S_0^r. Due to the time reversal invariance of the laws of classical mechanics, this microscopic description of the evolution of the system leads to the conclusion that the process is reversible.

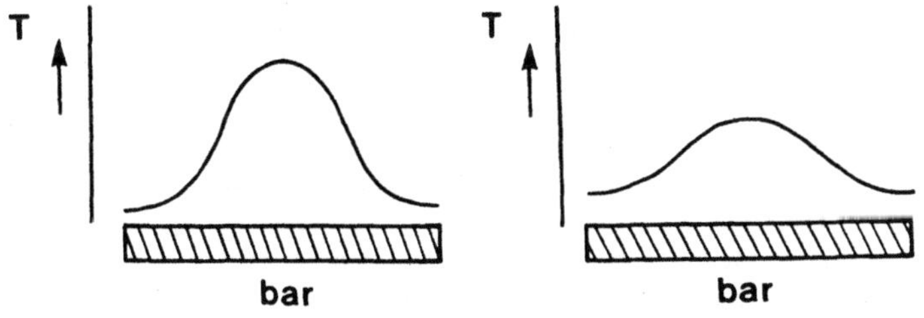

fig. 18a fig. 18b

Thus, if at a certain level of description the laws of nature violate time reversal invariance, then it remains possible that at another, more (less) fundamental level of description, time reversal invariance is not violated. In those cases, however, a reduction of one level of description to the other becomes problematic. With regard to the above example, a paradox arises as soon as it is assumed that the macroscopic description of the evolution of the bar is completely reducible to the microscopic description: how is it possible to reconcile macroscopic irreversibility with microscopic reversibility [Reichenbach 1971, p. 108 ff]? More about this problem will be said in the next chapter.

III.3.5 T*-violation in elementary particle physics.

Until about two decades ago, time reversal non invariant laws were known from the macroscopic, phenomenological level of description, but it was more or less generally assumed that at the microscopic level of description, the laws were time reversal invariant. This changed with the discovery of T*-violation in processes governed by the weak interaction force.

Roughly, the discovery of T*-violation in microscopic processes was the result of the following argument (11). Assuming that the laws governing the weak interaction obey certain very general conditions, it can be shown that, although the exact form of these laws is unknown, they must obey the CPT* theorem. Now, given the validity of the CPT* theorem for the weak interaction force, the T*-violation follows from the discovery that, in the decay of a neutral K meson, which is governed by the weak interaction, the PC symmetry is violated [Christenson 1964]. To save PCT* symmetry, it is therefore necessary that the T* symmetry be violated as well.

There is, however, no direct evidence for the time reversal non invariance of this decay process because the temporal reversal of this process, has (not yet) been studied in concreto. No direct comparison of the original process with its temporal reversal has been possible in practice. As of yet, only indirect evidence is available; the presumed T*-violation is based upon the assumptions that the weak interaction force obeys the CPT* theorem and that the interpretation of the CP violation in the decay process is correct. Yet, it has been possible to make a direct comparison for the cases of the strong nuclear and the electromagnetic force between processes and their time reversed processes, in the manner indicated in the quotation at p. 123. No violation of T* symmetry was found [Overseth 1969, p. 93 and 101].

The relevance of time reversal violation in elementary particle physics for the problem of time asymmetry, has been a point of controversy. If one is convinced that the only basis for time asymmetry is to be sought in the time reversal non invariance of the fundamental laws of physics, then the discovery of T*-violation in weak interaction processes is of capital importance, since here we encounter, for the first time, time reversal non invariant laws at the microscopic level; this could open the possibility of reducing macroscopic irreversibility to microscopic irreversibility. But as I stressed earlier, the existence of time asymmetry does not hinge exclusively upon the time reversal non invariance of physical laws. Moreover, it is hard to see how this T*-violation in weak interactions, which has only been observed in physical laboratories under highly artificial conditions, can account for the time asymmetric behaviour of macroscopic systems observed in everyday life. Therefore, the relevance of T*-violations in elementary particle physics for the problem of time asymmetry should not be overestimated; these T*-violations seem to be "unimportant outside the subnuclear domain" [Landsberg 1982, p. 6]. They do not contribute to a solution of the problem of the time asymmetric behaviour of systems observed in daily life. Any adequate theory of time asymmetry should in the first place solve that problem.

III.4 Time asymmetry and temporal orientability.

Remember that a spacetime manifold is temporally orientable if it is possible to distinguish globally in a coherent way between two opposite time orientations. A temporally orientable spacetime becomes temporally oriented by choosing one of the time orientations as the positive one; in principle, this choice is arbitrary (see p. 35-36).

At first, time asymmetry appears to require temporal orientability (or orientedness) of the spacetime manifold, for in the definition of time asymmetry, reference is made to two opposite time orientations. But a closer analysis shows that this is highly questionable. With regard to the connection between time asymmetry and temporal orientability (orientedness) of spacetime, I shall examine the following two questions: i) does temporal orientedness of a spacetime manifold imply time asymmetry, and ii) does time asymmetry require temporal orientability of spacetime?

ad i) A temporal orientable spacetime is transformed into a temporally oriented spacetime by arbitrarily choosing one of the two orientations of time as the positive or the future orientation. Thus, any temporally orientable spacetime, regardless of the occurrence of time asymmetric processes, can be transformed into a temporally oriented spacetime. Therefore, the temporal orientedness of a spacetime does not imply time asymmetry, and any attempt, like Earman's [1972a, 1974], to search for a basis for time asymmetry in that direction, is bound to fail (12). As Grünbaum remarks [1973, p. 790]:

> Time-orientability is merely a necessary condition for physically singling out the past and the future senses of time uniquely as such. For it assures only the globally consistent mere oppositeness of two senses of time by failing utterly to specify NON-trivially or non-arbitrarily which one of them is the future sense and which one is past!

If time asymmetry prevails, then the future facing orientation of time can be determined unambiguously and in a non-arbitrary way, by relating that future orientation to the asymmetry in the content of the world (e.g., mixing processes unambiguously single out the future orientation of time).

ad ii) Temporal orientability of spacetime is often considered to be a necessary condition for the possibility of the occurrence of time asymmetry. At first, this appears to be a sound position in view of the reference to two opposite time orientations in the definiton of time asymmetry. The problem, however, is whether this reference to two opposite time orientations is intended locally or globally. In the former case, time asymmetry presupposes temporal orientability of spacetime, whereas in the latter this is not necessarily so.

The issue can be put in a different way: is time asymmetry a local or a global phenomenon? At this point, it is necessary to recall that we have assumed that, in principle, time asymmetry has a regional character. Matthews [1979, p. 91-92] gives the following, strong argument for the regional character of time asymmetry, which concerns the assumption that the basis for time asymmetry in our actual universe is to be found in the prevailing boundary conditions:

> many of the attempts to define a direction of time [time asymmetry] do so on the basis of local boundary conditions together with the universal laws of physics... If these attempts are sound, then we should expect the physical phenomena which constitute the basis for a direction of time [time asymmetry] to be mani-

> fest in any region which possesses the appropriate boundary conditions; the laws of physics will automatically be satisfied in the region.

Thus he concludes [ibidem] that if time asymmetry is based on boundary conditions, "then we should expect it to be local or regional in character" [ibidem]. Indeed, the boundary conditions do not always and everywhere have to be the same in our universe (13).

Yet, it ought to be remarked that the regional character of time asymmetry does not imply that, from a topological point of view, time asymmetry is a local phenomenon! If, e.g., thermodynamic time asymmetry were a local phenomenon in the topological sense, then any arbitrary small neighborhood of a point of spacetime should exhibit thermodynamic time asymmetry; but this is clearly absurd, for there is a minimum size to the systems which can exhibit thermodynamic time asymmetry [Matthews 1979, p. 88]. An adequate treatment of the relation between time asymmetry and temporal orientability cannot avoid a discussion of the size of the systems exhibiting time asymmetry. Both 'small' and 'large' systems can behave asymmetrically in time; even the universe as a whole does exhibit time asymmetric behaviour. In case time asymmetry occurs with regard to systems which are small in comparison with the size of the universe, and for which the process time is short relatively to the time scale of the evolution of the universe, it does not appear to be necessary to assume temporal orientability for the whole spacetime; it would be sufficient to require temporal orientability for those regions of spacetime where time asymmetry occurs.

The situation with respect to the time asymmetric behaviour of the universe as a whole is more obscure; in that case, the global structure of the spacetime manifold, in particular the temporal orientability, appears to be much more relevant. Nevertheless, the question whether these 'global' time asymmetries are global in the proper topological sense is still open (14); therefore, even then the requirement of temporal orientability of spacetime can be put into question.

So far my attempt to clarify the concept of time asymmetry. Such clarification is an indispensible prerequisite for a discussion of the problem that will be the subject of the next chapter: what is the physical origin of thermodynamic time asymmetry?

NOTES

1 I have made no effort to trace back the origin of expressions like 'time asymmetry', 'anisotropy of time', 'orientation of time' and 'direction of time'. As is well known, the expression 'arrow of time' was introduced by Eddington in his Edinburgh Gifford lectures [Eddington 1928].

2 I am aware that the notion 'flow of time' is not without problems (see Epilogue); however I do not know what is to be gained by replacing it by Grünbaum's 'temporal becoming'.

3 It is not always possible to draw a clear boundary line between de facto and nomological irreversibility (for instance, in cosmology); cfr. Grünbaum [1973, p. 211].

4 This definition also agrees with the kind of time asymmetry encountered in elementary particle physics. The T*-violation in the subnuclear field leads to a difference in the rate with which processes and their reversal occur relatively to a given orientation of time [Landsberg 1982, p. 6].

5 A related problem presents itself with regard to the interpretation of statistical mechanics. As was pointed out to me by an anonymous referee, authors on statistical mechanics use by convention the ensemble language, but in principle they have a choice between using a probability language or an ensemble language.

6 It would have been more appropriate to refer to these processes as 'de facto irreversed' instead of as 'de facto irreversible'.

7 If m, in analogy to e, is considered to be a 'charge', then the cases $T^*(m) = m$ and $T^*(e) = e$ can be subsumed under the general rule that T* does not affect the sign of charges.

8 See Reichenbach [1971, p. 264 ff] and the references given there.

9 For more details on Stueckelberg's treatment of the second law of phenomenological thermodynamics, see sect. IV.3.1.

10 In quantum mechanics, for instance, the time reversal operator is given in the form of an anti-unitary transformation in contrast to the time-evolution operators which are given by unitary transformations [Davies 1977, p. 155-156].

11 For more details, see Davies [1977. p. 176] and Bunge [1968, p. 380-381].

12 Earman's position has been severely criticised by Grünbaum [1973, p. 788 ff].

13 Even if the cosmological principle is valid, irregularities can still arise due to local density fluctuations [Layzer 1971].

14 For more details on this point, and for the definition of global time asymmetries ('global' in the topological sense), see Matthews [1979].

CHAPTER IV

THERMODYNAMICAL TIME ASYMMETRY AND THE SECOND LAW OF PHENOMENOLOGICAL THERMODYNAMICS

IV.1 Introduction

IV.1.1 Different kinds of time asymmetry.

Time asymmetry pervades our whole universe, from the time asymmetric decay of the neutral K-meson to the expansion of the cosmos (1). In classifying the various kinds of time asymmetry, Penrose [1979] lists as many as seven types, viz., time asymmetry in K^0-meson decay, in quantum mechanical observations, in the general entropy increase, in the retardation of radiation, in psychological time, in the expansion of the universe and finally in black holes versus white holes. I shall briefly turn to each of them, whereafter I shall restrict myself to thermodynamical time asymmetry.

- K^0-meson decay. It has already been pointed out that the T*-violation in weak interaction decay is interesting because it concerns microscopic processes which behave asymmetrically in time and because this time asymmetric behaviour appears to be due to the laws governing these processes and not to boundary conditions. Yet, the effect of these time asymmetric decay processes is extremely small; the T*-violating component of the K^0-meson decay is only about 10^{-9} part of the T*-conserving component, and thus we are dealing here with a "very weak component of a weak interaction" [Penrose 1979, p. 583]. Its relevance for the macroscopic types of time asymmetry is very controversial (see sect. III.3.5).

- Quantum mechanical observation. In the standard interpretation of quantum mechanics, a system does not show any time asymmetric behaviour as long as its evolution is governed by the time reversal invariant Schrödinger equation. Only in the process of observation, i.e., of making a measurement, does the system go through an irreversible change, for then, the so-called 'collapse of the wave function' occurs. Suppose that, before the measurement, the system is in a 'pure' state;

in other words, its wave function ψ may be written as the superposition of the eigenstates corresponding to the observable A to be measured: $\psi = \Sigma_n a_n \phi_n$. After the measurement, the system will be in the eigenstate ϕ_j corresponding to the eigenvalue a_j, measured for the observable A. In the measurement process, the original state function $\Sigma_n a_n \phi_n$ is transformed into the state function ϕ_j; this is called the collapse of the wave function which is an irreversible process.

The precise status of the irreversible collapse of the wave function within quantum mechanics is rather obscure. The collapse of the wave function does not occur when the measuring apparatus is treated as an ordinary quantum mechanical system; in that case, the interaction between the measuring apparatus and the object system is governed by the Schrödinger equation and is therefore reversible. This 'measurement problem' is still one of the controversial issues in the foundations of quantum mechanics. Among other solutions, it has been suggested that in order to characterise a quantum mechanical interaction between two systems as an (irreversible) measurement process on one of these, it is necessary to make an appeal to a conscious observer who actually reads the result from the measuring apparatus (2).

- General entropy increase. Processes for which the entropy increases occur frequently in daily life; standard examples are the mixing of coffee and cream and the damped pendulum. This type of time asymmetric behaviour is described by the second law of phenomenological thermodynamics; if for a closed system the entropy increases, then the system undergoes an irreversible change. The status of this second law within physics is the main topic of this chapter.

- Retardation of radiation. Consider an antenna that produces electromanetic waves diverging into infinity. Such waves are called 'retarded radiation' because the electric currents in the antenna take place before the waves come into existence. The Maxwell equations, however, do not exclude the occurrence of the temporally reversed process: waves coming in from infinity, converging into the antenna and producing electric currents. Such imploding waves are called 'advanced radiation' because the waves exist before the electric currents take place in the antenna. The last situation is very unlikely to occur because very special boundary conditions must be satisfied in order to produce coherent incoming waves which converge into the antenna. Therefore, only retarded waves occur in our universe. Similarly, the time reversed process of the emission of light by stars does not occur; again, it requires very special boundary conditions to produce a 'star' which only absorbs radiation and thereby increases its energy content. In some cosmological models, this asymmetry between retarded and advanced radiation (between emission and absorption) is due to the expansion of the universe, whereas in others it is connected with thermodynamic time asymmetry [Davies 1977, Ch.V].

- Psychological time. In his survey of the different kinds of time asymmetry, Penrose [1979, p. 591] describes the asymmetry in psychological time as "the feeling of relentless forward temporal progression, according to which potentialities seem to be transformed into actualities." But this is what we have been referring to as the flow of time. Therefore, this characterisation of the asymmetry of psychological time has to be rejected since it obscures the difference between time asymmetry and the flow of time, which was introduced in the previous chapter.

It is possible to extend the notion of time asymmetry from the description of physical reality to the domain of subjective experience so as to avoid any reference to the flow of time. Psychological time is asymmetric in the following sense: the information stored in my memory increases relatively to one orientation of time, and decreases relatively to the other. Therefore, with regard to my personal experience, both orientations of time are not equivalent.

- The expansion of the universe. The expansion of the universe is usually referred to as the 'cosmological arrow of time'. There is still no conclusive evidence that the current expansion will continue indefinitely; it may be followed by a period of contraction in which case the cosmological arrow would be reversed. In discussions about time asymmetry, the expansion of the universe often plays an important role because it is considered to be the origin of other types of time asymmetry (e.g., the asymmetry in retarded and advanced radiation).

- Black holes versus white holes. The relevance of 'black holes' for the problem of time asymmetry is still very controversial. There are some indications that, from a thermodynamical point of view, black holes behave according to the second law of phenomenological thermodynamics - if it is assumed that the entropy of a black hole is proportional to the surface O of the black hole's absolute event horizon, the increase of its entropy follows immediately from the fact that O cannot decrease [Davies 1977, p. 97-102]. According to Penrose, the existence of the time reversed counterpart of black holes, i.e. white holes, cannot be ruled out since the laws of general relativity are time reversal invariant. These white holes will show an anti-thermodynamic behaviour: their entropy will decrease. As in the case of advanced radiation, attempts have been made to rule out the actual occurrence of white holes in our universe on the grounds that the relevant boundary conditions for producing white holes are extremely improbable [Penrose 1979, p. 600-611].

Apart from the problems connected with each type of time asymmetry separately, the existence of different kinds of time asymmetry raises the important question whether or not they are related to each other; in other words, whether or not these different time asymmetries have a common origin. The unraveling of the connections between them is an extremely complicated matter; a better insight into these issues will depend particularly on further developments in cosmology (3).

IV.1.2 The origin of thermodynamic time asymmetry.

It is rather asthonishing that although almost all phenomena of the world are clearly irreversible, physics nevertheless has managed so very well without taking into account this irreversibility in its basic theories. The laws of classical, relativistic and quantum mechanics, the laws governing the gravitational, the electromagnetic, the strong and the weak interaction (except for some T*-violating processes governed by the weak interaction), all are or are supposed to be time reversal invariant. Whereas almost all processes observed in daily life happen 'in just one way' and therefore allow us to distinguish unambiguously between the past and future orientation of time, these fundamental laws of physics are the same for both orientations.

This brings us to the central theme of this chapter, namely the relation between time asymmetry and the laws of physics. Is it possible to account for the observed kinds of time asymmetry in our universe in terms of the time reversal invariant laws of physics mentioned above, or is the irreversible character of the phenomena so fundamental that the introduction of time reversal non invariant laws into the basic theories of physics cannot be avoided? In discussing this problem, I shall confine myself to thermodynamic time asymmetry because here the contrast between a reversible microscopic description and an irreversible macroscopic description shows itself most clearly (see p. 129 ff).

Consider once more Feynman's example of the mixing of blue ink with water. After having remarked that at the molecular level all the collisions between the molecules of water and ink are perfectly reversible, and that therefore, after mixing, the ink and water could be separated again, he writes [1973, p. 112]:

> Things are irreversible only in a sense that going one way is likely, but going the other way, although it is possible and is according to the laws of physics, would not happen in a million years. It is just ridiculous to expect that if you sit there long enough the jiggling of the atoms will separate a uniform mixture of ink and water into ink on one side and water on the other.

And thus he concludes that [op. cit. p. 114]:

> the apparent irreversibility of nature does not come from the irreversibility of the fundamental physical laws; it comes from the characteristic that if you start with an ordered system, and have the irregularities of nature, the bouncing of the molecules, then the thing goes one way.

According to this widespread point of view, irreversibility in nature does not originate from the laws of physics, but is caused by the "general accidents of life" [op. cit. p. 112] and is therefore a de facto irreversibility.

Nevertheless, processes like the mixing of blue ink with water leave the impression that they behave irreversibly in a strictly lawlike way. The mixing of both components appears to be a necessary consequence of eliminating their separation. In other words, it looks as if a time reversal non invariant law which distinguishes between past and future orientation of time is governing these kinds of processes. The second law of phenomenological thermodynamics is such a time reversal non invariant law (4), and it adequately describes the irreversible behaviour of many macroscopic systems. It states that, relatively to the future orientation of time, the entropy of any closed system increases or remains equal, but never decreases; it is universally valid within a phenomenological description and can therefore appropriately be qualified as a fundamental law of nature (5). Thus, within a macroscopic or phenomenological description, it appears that irreversibility finds its origin not in the accidents of life but in one of the basic laws of physics.

But precisely, how 'basic' is this second law of phenomenological thermodynamics, especially in comparison with the laws of classical mechanics (6)? Is it as fundamental as these time reversal invariant laws, or is it only a 'second rate' law which, although adequately describing the behaviour of macroscopic systems, cannot be valid because, as we will see, it is incompatible with classical mechanics? Here we come to the core of our problem. These questions deserve close scrutiny since their answers determine to a large degree the physicist's conception of thermodynamic time asymmetry. It leads either to the idea that irreversibility can be traced back directly to one of the fundamental laws of nature, or to the point of view that its origin lies elsewhere, viz., in the boundary conditions prevailing in our universe, and therefore is but one of its accidental features.

Ever since the problem of the 'arrow of time' has surfaced, both points of view have been defended by physicists. This has led to two competing research programs or traditions with regard to the problem of the origin of thermodynamic time asymmetry: the mechanical and the thermodynamical program. Both have the same goal, namely to provide an explanation of thermodynamic time asymmetry as observed in daily life; but their approach to the problem is different. Quite generally, the mechanical program prefers the reversible mechanical description of physical reality to the irreversible thermodynamical one, whereas the thermodynamic program takes the opposite point of view: the irreversible thermodynamic description is considered to be more basic than, or at least the equal of the reversible mechanical description.

- The mechanical program. The point of view underlying the mechanical program has become standard within physics. This program finds its origin in the work of the famous Austrian physicist Ludwig Boltzmann [1964], dating from the end of the 19th century; in the 20th century, important contributions to a further development of this program were made, among others, by the Ehrenfests [1959], Reichenbach [1971] and Grünbaum [1973]. According to this position, thermodynamic time asymmetry is, relative to the most fundamental, mechanical level of description, of a de facto nature since the laws of mechanics are time reversal invariant. Moreover, the validity of the second law of phenomenological thermodynamics in its original form is questioned because it is incompatible with the laws of mechanics.

The origin of thermodynamic time asymmetry is sought in the boundary conditions prevailing in our universe. These must be such that, in combination with the laws of mechanics, they can explain the lawlike irreversible behaviour of macroscopic systems as observed in daily life. The most famous attempt in this direction is Reichenbach's 'branch systems' approach which will be examined later on. Schematically, this explanation of thermodynamic time asymmetry has the following structure:

Reversible laws of microscopic physics
Appropriate boundary conditions

Lawlike irreversible behaviour of macroscopic systems.

In other words, the fundamental laws of physics taken together with some specific boundary conditions can give rise to less fundamental laws, such as the second law which adequately describe processes behaving irreversibly in a lawlike manner. Several proposals for the specific nature of these boundary conditions have been put forward, such as the fact that the initial state of the universe was a state of relatively low entropy, the dissymmetry between matter and antimatter, the expansion of the universe, and Reichenbach's branch systems hypothesis. In this program it is therefore assumed that the paradox between macroscopic irreversibility and microscopic reversibility (see p. 131) can be resolved without tampering with the laws of classical mechanics.

- The thermodynamical program. Advocates of this point of view are relatively scarce among physicists, and in comparison with the mechanical program, there is far less continuity in the development of this position. It has been defended by, e.g., Max Planck who for some time thought that by giving up the hypothesis of the atomicity of matter, it would be possible to reconcile the second law of phenomenological thermodynamics with the laws of classical mechanics. For him, this second law had the same 'absolute' status as classical mechanics [Kuhn 1978, p. 23-25]. More recently, Stueckelberg [1974] and Prigogine [1980] have defended the thermodynamical program; their ideas will be discussed extensively later on.

Phenomenological thermodynamics with its time reversal non invariant second law constitutes the 'hard core' of this program (7). In view of the incompatibility of the second law with classical mechanics, the foregoing implies that the validity of classical mechanics must somehow be restricted or that classical mechanics must be embedded coherently into a broader framework in which there is room for time reversal non invariant laws. This, of course, is one of the main problems of the thermodynamical program: how to reconstruct classical mechanics so as to make it compatible with the second law, but without causing the collapse of the whole building of theoretical physics? In view of this danger, it is not surprising that advocates of this program are rather scarce.

My intention in this chapter is to give a detailed exposition of both programs, and particularly of the thermodynamical one, since it is usually neglected in the literature on thermodynamic time asymmetry, and to compare the two. More specifically, I shall extensively discuss the recent attack on the mechanical program by Prigogine in his book _From being to becoming_.

IV.2 The mechanical program

From the point of view of the mechanical program, the history of the second law of phenomenological thermodynamics is a history of attempts to reconcile this law with classical mechanics, that is, to explain its time reversal non invariance on the basis of the time reversal invariant laws of mechanics, or to reformulate it such that it becomes compatible with those laws. It is impossible to describe these attempts in a few pages since such description would have to include an exposition of the evolution of phenomenological thermodynamics via the kinetic theory of gases into statistical mechanics [see, for instance, Brush 1976]. Therefore, I shall confine myself to a schematic exposition of i) Boltzmann's deterministic H-theorem for individual systems, which he thought to be equivalent to the second law, and the reversibility objections raised against this theorem, and ii) the statistical mechanical H-theorem which is not subject to these reversibility objections. This statistical H-theorem is generally regarded to be the reformulation of the second law of phenomenological thermodynamics into statistical mechanics. On the basis of this H-theorem, it became clear that the entropic behaviour of an isolated gas does not exhibit any time asymmetry, since the statistical H-theorem is symmetric with regard to the two orientations of time. Thus, within the mechanical program, the time asymmetric behaviour of thermodynamical systems cannot be traced back to this H-theorem; its origin has to be sought elsewhere. Different solutions have been proposed, of which I shall discuss only the most important ones, viz., the attempts to explain thermodynamic time asymmetry on the basis of i) certain limitations in human observations and ii) the behaviour of (ensembles of) open systems.

IV.2.1 Boltzmann's deterministic H-theorem and the reversibility objections.

Boltzmann's aim was to recover the second law of phenomenological thermodynamics within the context of the kinetic theory of gases; this theory is based upon classical mechanics and the molecular or atomistic conception of matter (8). But he was forced to introduce an additonal assumption about the behaviour of molecules in a gas, known as the assumption of molecular chaos (the 'Stosszahlansatz') [Boltzmann 1964, p. 40]. Although this assumption can be formulated mathematically in a very precise way - it is expressed by the equality of two distribution functions - its physical interpretation and implications are not well understood [Davies 1977, p. 57]. Usually, the assumption of molecular chaos is interpreted as stating that there is no correlation between the velocities and positions of the molecules of the gas before they collide. This appears to be a rather plausible assumption if the molecules of the gas are considered to be moving around randomly. Boltzmann [1964, p. 40] gives the following examples of gases for which the assumption is not valid:

> We have a molecular-ordered distribution if - to select only two examples from the infinite manifold of possible cases - each molecule is moving toward its nearest neighbor, or again if each molecule whose velocity lies between certain limits has ten much slower molecules as nearest neighbors.

I shall not go further into the problem of the interpretation of molecular chaos.

Assuming that molecular chaos prevails at each moment during the evolution of a gas, Boltzmann was able to define a quantity H which is a function of the state of the gas, that is of the position and momentum of the molecules of the gas, so that:

$$\frac{dH}{dt} \leq 0,$$

with $H(t) = H(x_i(t),p_i(t))$. In Boltzmann's opinion, the second law could be obtained immediately from this inequality by equating H with minus S, and thus by making entropy a function of the phase space variables x_i and p_i.

Boltzmann originally interpreted this H-theorem in a deterministic way: for any gas in whatever state it would be possible to define a quantity H which in the course of the evolution of the gas would decrease or remain equal, but would never increase. But he failed to see that his derivation of the H-theorem did not allow for such deterministic interpretation since he had made use of probabilistic methods in deriving this result [Kuhn 1978, p. 44]. Moreover, another problem presented itself. The quantity H as defined by Boltzmann, does not change sign under time reversal, that is under a replacement of p_i by $-p_i$; consequently, the H-theorem is not time reversal invariant. But nevertheless, its derivation was based upon classical mechanics; starting from the time reversal invariant laws of classical mechanics, Boltzmann claimed to have deduced a time reversal non invariant law. This gave rise to the so-called 'reversibility objections' (10).

i) Loschmidt and others objected that the deterministic H-theorem could not be correct because it was based upon the assumption that the motions of, and the collisions between the molecules of the gas were reversible. How, then, would it ever be possible to arrive at a strictly irreversible behaviour of a gas as predicted by the H-theorem? More particularly, since H is a function of x_i and the absolute value of p_i, it follows that if H decreases in the transition from state (x_i^0,p_i^0) to state (x_i^1,p_i^1), then H must necessarily increase in the equally possible temporally reversed process from $(x_i^1,-p_i^1)$ to $(x_i^0,-p_i^0)$. Therefore, any decrease of H in a process A, necessarily implies the increase of H in the temporally reversed process A^r which is also allowed by the laws of physics. Thus, the H-theorem could not be valid.

ii) In his 'recurrence theorem', Poincaré had shown that an isolated system passes through a sequence of states so that the system will return, an infinite number of times, to an arbitrarily small neighbourhood of any one of its states. Starting from this result, Zermelo also objected that the H-theorem could not be valid; if the recurrence theorem of Poincaré holds, then this implies that if, somewhere along the trajectory representing the evolution of an isolated system, H decreases, then it has to increase elsewhere along the trajectory.

From these objections, it became clear that something had to be wrong with the assumption of molecular chaos. And indeed, it can be shown that a contradiction arises between classical mechanics and the assumption that molecular chaos prevails at each moment during the evolution of a gas, as was assumed by Boltzmann [Davies 1977, p. 58]. This assumption therefore had to be abandoned.

IV.2.2 The time symmetrical statistical H-theorem.

Statistical mechanics avoids these reversibility objections by defining a statistical analogue of the entropy function and of the second law. This statistical approach differs strongly from the one followed originally by Boltzmann; no attempt is made to calculate in a deterministic way the behaviour of an individual system represented by a swarm of points in μ-space (μ-space is the six-dimensional space with coordinates x,y,z and p_x,p_y,p_z; a point in μ-space represents one molecule of a gas at a definite place with a definite momentum). Instead, the most probable behaviour of the system is computed by starting from an ensemble of similar systems, represented by a swarm of points in Γ-space (Γ-space is a 6N-dimensional space with N the number of molecules in a gas; a point in Γ-space represents the complete microscopic state of one of these systems, i.e., contains complete information about places and momenta of all the molecules of the gas). Accordingly, this approach does not allow to make predictions about the actual behaviour of an individual system, but only about its most probable behaviour (or about the mean behaviour of an ensemble of systems) (9).

Roughly, the statistical analogue of Boltzmann's deterministic H-theorem states that, given an ensemble of systems not in equilibrium, the value of H will decrease for the vast majority of these systems, and that H will increase for a small number of systems only (11). In other words, the most probable behaviour of a system not in equilibrium is toward equilibrium, where H (the entropy S) has a minimum (maximum) value. In this version, the H-theorem is insensitive to the reversibility objections because it does not say anything about the actual behaviour of a single system. For an isolated mechanical system, the reversibility objection of Zermelo, for instance, implies that, after an initial decrease, H eventually must increase; but this fact does not contradict the statistical H-theorem which only says something about the most probable behaviour of a system not in equilibrium [Davies 1977, p. 56].

According to the statistical mechanical interpretation of the entropy concept, the entropy of an isolated gas is not constant, as was the case for phenomenological thermodynamics, but fluctuations away from the maximum value of S occur due to small deviations from the equilibrium state of the gas. This is illustrated in fig. 19. This conception of entropy is usually referred to as the 'Boltzmann entropy' (12). Because the number of microstates corresponding to the macroscopic equilibrium state (the state of maximum entropy) is overwhelmingly larger than the number of microstates corresponding to non-equilibrium states with relatively low entropy values, the gas will mostly be in a state of (almost) maximum entropy. Strong deviations from equilibrium are very improbable (see also sect. IV.4).

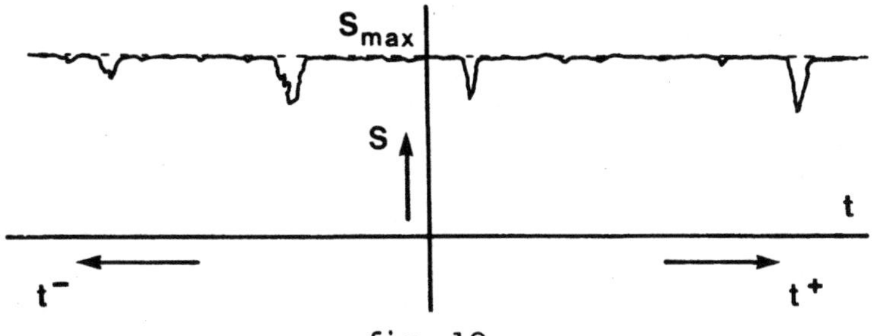

fig. 19

The question to be considered now is whether or not this statistical H-theorem leads to an asymmetric behaviour of an isolated gas. In this respect, the most fundamental result of the statistical mechanical reformulation of the second law of phenomenological thermodynamics is that the entropic behaviour of an isolated gas does not show any sign of time asymmetry. Fig. 19, showing the entropy curve of such system, helps to understand why this is the case: there is no difference between the entropic behaviour of an isolated gas as seen relative to the future (t^+) or the past (t^-) orientation of time. Relative to the past as well as the future orientation of time, the entropy increases as often as it decreases. Thus, whenever a gas is in a relatively low state of entropy at moment t_0 and in a state of maximum entropy at t_1, we are not allowed to conclude that t_0 is earlier than t_1. Since transitions from low entropy to high entropy states occur as frequently as the reverse transitions relative to both orientations of time, we cannot single out unambiguously one of these orientations as the future orientation on the basis of the entropic behaviour of a closed gas.

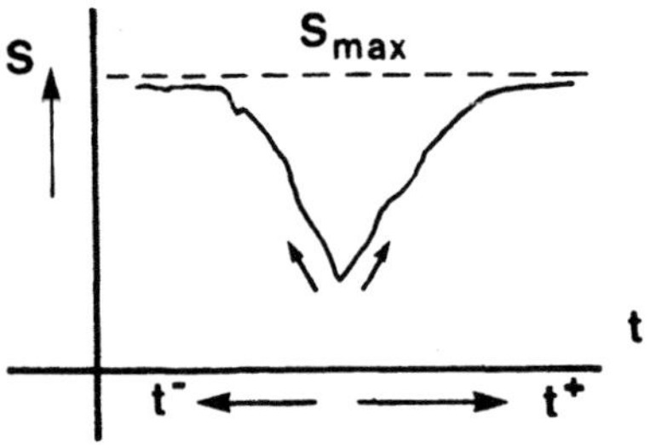

fig. 20a

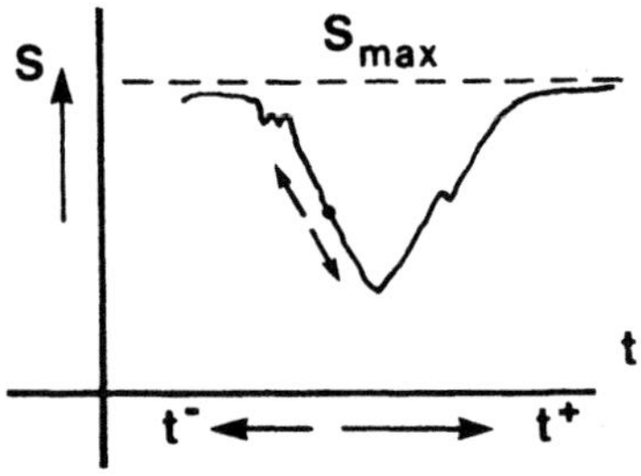

fig. 20b

Nevertheless, the statistical mechanical H theorem is valid: the most probable behaviour of a gas not in equilibrium is toward equilibrium. This can be seen as follows. When an isolated gas is observed to be in a relatively low state of entropy, then the system is most probably at or near the bottom of a downgrade fluctuation (see fig. 20a), since it is rather unlikely that the gas is on the downgrade of a still stronger deviation from equilibrium (see fig. 20b) (13). But from the bottom of a downgrade fluctuation, the entropy increases relative to both orientations of time, as predicted by the statistical H-theorem. Thus, the statistical H-theorem does not introduce any time asymmetry in statistical mechanics and is compatible with the time symmetric behaviour of an isolated gas. That was to be expected in view of the fact that this theorem is derived within the framework of classical mechanics.

In spite of the symmetry in time of the statistical H-theorem, we observe in daily life only systems which by themselves develop from low entropy states to high entropy states; never the reverse. How can this be made plausible within the mechanical program?

Various kinds of solutions have been proposed; I shall restrict myself to a discussion of 1) solutions in which certain limitations inherent to human observation play a dominant role, and 2) solutions in which, in one way or another, the assumption that the system is (perfectly) closed is abandoned. In the latter case, the interaction of the system with its environment is considered to be the origin of its time asymmetric behaviour. The most famous theory starting from this point of view is Reichenbach's 'branch systems' theory; it deserves our special attention, since it represents the mechanical program in its most sophisticated form.

IV.2.3 Thermodynamic time asymmetry and human observation.

Two specific limitations of human observation have been connected with thermodynamic irreversibility, namely, i) the relatively short span of time involved in human observation, and ii) the lack of microscopic information in human observation.

i) The time interval Δt during which a human being can observe the behaviour of a closed macroscopic system is extremely small compared with its typical Poincaré recurrence time T (T is roughly the mean time in which the closed system returns, or almost returns, to its original state) (14). This fact is sometimes linked to the time asymmetric behaviour of a non equilibrium closed system: this asymmetry would disappear if we would observe the system long enough, that is if Δt becomes of the order of T [Earman 1974, p. 38]. But this relatively short observation time Δt cannot explain that we only observe transitions from non equilibrium to equilibrium, but not the reverse. Consider again the entropy curve of an isolated gas as given in fig. 19. If we observe the evolution of the gas through a 'slit' of Δt sec., assuming that the slit is placed randomly somewhere along the time axis, then, if we repeat the experiment many times, transitions from high to low entropy will be observed as frequently as the reversed transitions, regardles of whether this slit is small or large relatively to the recurrence time T of the system. No asymmetry will be observed. Of course, if the slits are not distributed at random along the time axis, then the illusion that the system behaves asymmetrically in time, might arise. For example, if the beginning of the slit is always situated at the bottom of a downgrade fluctuation, then a transition from low to high entropy states will be observed in most cases. But it will be clear that this seemingly time asymmetric behaviour of the gas is not an objective feature of the evolution of the system itself, but originates from the peculiar conditions under which the system is observed.

ii) Because of our 'poor' means of observation, it is impossible to exactly locate the points representing the different molecules in phase space. This means that, in describing the evolution of the gas, we are forced to use methods like 'coarse graining of phase space', that is, phase space has to be divided into finite compartments or cells, and the only information available through human observation is how the various molecules are distributed over these cells. But precisely with the help of techniques like coarse graining, it is possible to define a 'coarse grained entropy' which, for a system not in equilibrium, most likely increases in time; this stands in sharp contrast with the 'fine grained entropy' which is constant in time and which is based upon the exact knowledge of places and momenta of all the molecules [Davies 1977, p. 49 ff]. Sometimes, the conclusion drawn from this is that if complete information about the system were available, the seemingly time asymmetric behaviour of the system would disappear. According to this point of view, the basis for thermodynamic time asymmetry lies in our ignorance of the precise motions of all the molecules of the system, in other words, in our lack of microscopic information [Layzer 1972, 1975].

A problem with this position is that it explains thermodynamic time asymmetry and irreversibility in terms of anthropomorphic concepts such as human ignorance. The same objection of anthropomorphism has been leveled against the use of the concept of entropy, because the definition of entropy also depends upon the process of coarse graining phase space (15). To avoid this criticism, Layzer [1972, p. 281] has tried to distinguish objectively between macroscopic and microscopic information, in order to give an objective meaning to expressions such as "lack of microscopic information". I shall not consider the question whether or not this problem of anthropomorphism can be resolved, since in my opinion there is a much stronger reason to doubt the validity of this kind of explanation of thermodynamic time asymmetry. The whole line of reasoning is based upon a false premise, namely that the use of methods like coarse graining in statistical mechanics, regardless of the presence or lack of microscopic information, leads to a time asymmetric description of the evolution of a closed gas. This, however, is not the case. According to the statistical H-theorem which is based upon coarse graining of phase space, a closed system behaves perfectly symmetrically in time, as we have seen before.

Undoubtedly, these attempts to connect thermodynamic time asymmetry with limitations in human observation account for, or have strengthened the idea that within the mechanical program the only possible solution for the conflict between the reversible laws of physics and the observed time asymmetric behaviour of thermodynamic systems lies in degrading the latter into an illusion. According to this idea, adherence to the mechanical program implies that thermodynamic time asymmetry cannot be considered to be real, i.e., an aspect of physical reality, but necessarily leads to the point of view that thermodynamic time asymmetry is an illusion caused by certain peculiarities of human observation. I shall return to this issue in section IV.4.

IV.2.4 Thermodynamic time asymmetry and open systems.

So far, we have assumed that the systems involved are perfectly closed; now, we shall assume that there is some interaction between the system and its environment. In discussing the relevance of the openess of the system for the problem of thermodynamic time asymmetry, I shall distinguish between two types of interaction between the system and its environment:

i) The interaction is very weak so that it does not destroy the system nor alters any of its specific features; in other words, the system is not perfectly closed but quasi-isolated. In practice, this is always the case for any allegedly closed system; for instance, it is impossible to shield off the gravitational interaction from a system.

ii) The interaction is such that at a certain moment a quasi-isolated system is formed, which eventually merges back into its environment. Here, the interaction is responsible for creating the system and, possibly, also for its destruction; meanwhile, the system behaves as a quasi-isolated system. As an example, consider an ice cube in a glass of water [Reichenbach, 1971, p. 118]; the system is created the moment the ice cube is put in the water. During the melting of the ice cube, the system can be considered to be isolated "though the thermic isolation is not too good" [ibidem]. Eventually, the system will merge, from an entropic point of view, into its environment since it will reach thermodynamic equilibrium with that environment.

In case the interaction of the system with its environment is very weak, the interaction can be regarded as a perturbation of the system. Let us assume that these perturbations take place at random. For a gas, these random perturbations will tend to destroy at each moment the correlations between the molecules of the gas, and therefore, they will tend to restore the situation of molecular chaos at each moment. Note that even if they destroy all the correlations and thus maintain molecular chaos during the entire evolution of the gas, Boltzmann's deterministic H-theorem would not be valid, because Boltzmann overlooked the probabilistic element in what he took to be a deterministic H-theorem.

The role of random perturbations in establishing equilibrium should not be overestimated, for the "internal relaxation time - the time required for the gas to reach internal equilibrium - is typically 10^4 times shorter than the time required to reach equilibrium with the walls" [Davies 1977, p. 75]. Thus, the internal collisions, not those with the walls, cause a gas to approach equilibrium. Consequently, the influence of random perturbations operating on the gas by way of the walls, is very small, if not neglegible. Moreover, these random perturbations will result a priori as often in deviations from as in returns to equilibrium. Therefore, they cannot introduce any time asymmetry in the behaviour of a quasi-isolated gas. The entropy curve of such system will be of the same form as the one given in fig. 19; to any deviation from equilibrium there corresponds an approach to equilibrium.

The gravitational interaction poses a special problem. Although it is the weakest of all known interactions, it does influence the microscopic evolution of a gas considerably , since "the change in gravitational potential caused by displacing one gram of matter by one centimeter at the distance of the star Sirius would, in the course of one microsecond, substantially alter the microscopic state of a macroscopic volume of gas" [Layzer 1975, p. 66]. The influence of the gravitational interaction upon the way a gas approaches equilibrium, is not known. But whatever its effect on the evolution of a gas, it appears that the gravitational interaction, considered as random perturbations from the environment, cannot account, for the reasons given above, for thermodynamic time asymmetry (16).

I shall now turn to Reichenbach's branch systems approach in which the second kind of interaction mentioned above plays a fundamental role. His point of view was adopted by Grünbaum [1973, p. 209 ff], albeit with some modifications. Instead of concentrating upon the behaviour of one single permanently closed system, Reichenbach analyses the statistical behaviour of an ensemble of isolated systems which exist only for a short span of time. These systems are called 'branch systems'; they "branch off from a comprehensive system and remain isolated from then on for some length of time" [Reichenbach 1971, p. 118]. Under certain conditions, an ensemble of branch systems does exhibit a statistical asymmetry in time. Given an ensemble of branch systems, each of which at the moment of formation t_0 is in a relatively low state of entropy, and given the assumption that most of these systems are at the bottom of a dip in the entropy curve, it follows that after some time the majority of these branch systems will be in a relatively higher state of entropy. But it does not follow that, likewise, some time before t_0, the majority of these branch systems were in a relatively high state of entropy: indeed, they did not exist. Of all the branch systems, the future upgrades to equilibrium are left untouched, whereas the past upgrades to equilibrium, which are reponsible for the symmetric behaviour of a permanently closed system, are cut off because the branch systems were not yet formed (see fig. 21). Therefore, it is wrong to draw the conclusion that, some time before t_0, the majority of the branch systems was in a state of relatively high entropy (17).

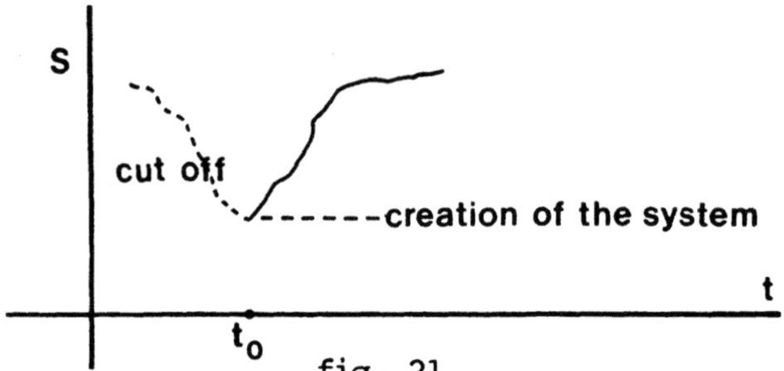

fig. 21

Not every ensemble of branch systems exhibits a time asymmetric behaviour. The requirement that these branch systems, at the moment of their creation, be in a relatively low state of entropy is essential to Reichenbach's argument. An ensemble of branch systems, all of which are in equilibrium at the moment of their creation, does not exhibit any time asymmetry.

In order to apply his analysis of the behaviour of ensembles of branch systems to the real world, Reichenbach must introduce an assumption called the 'branch systems hypothesis' [1971, p. 135 ff]. This assumption must assure the existence in nature of the branch systems of the required type. Roughly, the branch systems hypothesis states that

1) during the current cosmic epoch, the universe is in a relatively low state of entropy, and

2) within our region of the universe, a large number of closed or quasi-closed systems in a relatively low state of entropy branch off constantly from the universe as a whole.

Thus, Reichenbach assumes that nature abounds in branch systems which start off in a relatively low state of entropy, and he is convinced that this assumption can be regarded as empirically verified. Given this assumption, it is now possible to define objectively the future orientation of time in terms of the time asymmetric behaviour of these branch systems: "The direction in which most thermodynamical processes in isolated systems occur is the direction of positive time" [op. cit. p. 127]. The future orientation of time coincides with the orientation relatively to which the entropy of the universe as a whole increases; through the continual creation of branch systems, this entropy increase of the universe becomes apparent to us [op. cit. p. 131]:

> The existence of a long upgrade of entropy [of the universe], though a necessary condition for the phenomenon of time direction, is therefore not a sufficient condition. Time direction becomes apparent to us only because the upgrade contains a large number of situations in which subsystems branch off, disclosing in their further development the universal growth of entropy. This growth would not become evident unless the entropy of the subsystem initially had a low value as compared with the entropies of other possible states of the subsystem while it remains isolated. Our universe satifies this condition. It owes its time direction, therefore, to the branch structure of the upgrade of its entropy curve on which we live.

Thus, the general entropy increase in the universe as a whole is reflected in the thermodynamic behaviour of the majority of branch systems; only through them, does the general increase of entropy become observable to us.

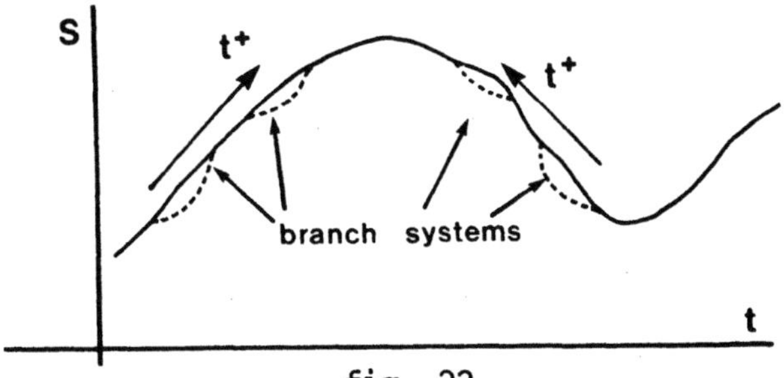

fig. 22

According to Reichenbach [op. cit. p. 127], the entropy of the universe is subject to fluctuations (see fig. 22). The application of his definition of the future orientation of time to an upgrade and a downgrade of the entropy curve of the universe shows that the resulting future orientations of time are counter-directed. Therefore he remarks that "we cannot speak of a direction for time as a whole; only certain sections of time have directions, and these directions are not the same" [ibidem].

On this point, he was criticised by Grünbaum who objected against the use of he notion of the entropy of the universe as a whole, because this notion is not well defined [1973, p. 261-262]. The details of Grünbaum's criticism are not directly relevant within the present context since they do not concern in any way the most essential aspects of Reichenbach's scheme of explanation, viz., that the origin of thermodynamic time asymmetry lies in the prevailing boundary conditions in our universe. For both Reichenbach and Grünbaum, boundary conditions plus the time reversal invariant laws of physics, more particularly of classical statistical mechanics, can in principle explain the observed time asymmetric behaviour of thermodynamic systems.

To this day, the branch systems approach is one of the most sophisticated attempts within the mechanical program to base thermodynamic time asymmetry on boundary conditions. Returning to the scheme of p. 145, the goal of the mechanical program seems very nearly to have been reached. The time asymmetric behaviour of low entropy branch systems can indeed be deduced from the boundary conditions specified in the branch structure hypothesis (18), together with the time symmetric statistical H-theorem. It is also possible to understand why 'counter-thermodynamic' behaviour, from equilibrium to non-equilibrium, is so extremely improbable that it de facto never occurs; that explains the lawlike character of the irreversible approach to equilibrium (see also sect. IV.4). Nevertheless, the branch systems approach in its present form shows some shortcomings which will be discussed in the final section.

IV.3 The thermodynamical program

In the mechanical program, the contradiction between the second law of phenomenological thermodynamics and mechanics is removed by reformulating and reinterpreting the second law. I shall now consider some views which, generally, follow the opposite strategy: they start off from the validity of the second law of phenomenological thermodynamics, and they assume that, somehow, classical mechanics has to be reinterpreted and reformulated so that it becomes compatible with the second law. According to these views, the roots of thermodynamic time asymmetry lie in this basic law of nature and not in some boundary conditions.

Advocates of this point of view are rather few among physicists because a commitment to this line of thought makes it necessary to reconsider and, eventually, to reconstruct one of the main pillars of theoretical physics. But even though the defenders of the thermodynamical program are a minority, it is worthwhile to examine their ideas since they have far-reaching consequences for the physicist's conception of thermodynamic time asymmetry. The most important contemporary representatives of this non-standard view are Stueckelberg and Prigogine. Both try, albeit in different ways, to forge phenomenological thermodynamics and mechanics into a unique scheme which harmoniously embraces both the thermodynamical and the mechanical mode of description.

IV.3.1 Mechanics as a special case of thermodynamics.

In Thermocinétique phénoménologique galiléenne, Stueckelberg and Scheurer expose the idea that the study of physical reality ought to start with and to proceed from phenomenological thermodynamics (19). Instead of following the traditional path of commencing the study of the physical world by 'rational mechanics' ('la mécanique rationelle'), in which the notion of a material point particle is introduced by way of an axiom, Stueckelberg begins with the part of physics that is usually referred to as 'phenomenological' [1974, p. 9]. He opposes this phenomenological physics, of which the spacetime continuum such as it appears in observations of macroscopic material systems is an essential aspect, to microscopic or quantum physics in which discontinuous processes play a fundamental role. Within this phenomenological description, fluctuation phenomena can be discarded and therefore there is no need, according to Stueckelberg, to resort to the methods of statistical mechanics.

Time plays an important role within phenomenology. The succession of events leads to the conception of time as a one-dimensional and totally ordered continuum. But time has yet another, additional feature, viz. the direction of time ('la flèche du temps'), which accounts for the dissymmetry between past and future and the irreversible evolution of the phenomena. This essential aspect of time destroys Laplacean determinism, according to which the state of an isolated system at any time determines completely the state of the system at all other times, in the past as well as in the future. This Laplacean determinism is one of the most salient features of the mechanistic worldview; once the positions and the momenta of the particles of a system are known, then its whole history and also its whole future are in principle determined. But within a phenomenological description this is precisely impossible. The symmetry between past and future is broken by the fact that the solutions of the differential equations describing the evolution of a system converge in the distant future but diverge in the distant past because of the signs of the constants in the equations of motion (like friction coefficient or heat capacity) [op. cit. p. 10]. Usually, the signs of these constants are fixed afterwards so that the solutions obtained converge toward the future; but Stueckelberg shows that their signs are determined by the second law of phenomenological thermodynamics. Thus, this second law is incompatible with the idea of Laplacean determinism; it implies that it is possible, in principle, to predict the whole future of an isolated system, whereas it is not possible to reconstruct its entire past on account of the occurrence of a singularity within a finite span of time (as in the case of the 'Big Bang' theories) (21).

Because irreversibility is such an essential aspect of the phenomena, and because this irreversibility is intimately connected with the increase of entropy, Stueckelberg is led to the conclusion that the study of the physical world is best started by that part of phenomenological physics which is ordinarily called 'phenomenological thermodynamics', but which he prefers to call 'thermocinétique phénoménologique', because forces do not play an essential role in the equations of motion. It is not Stueckelberg's intention to oppose this phenomenological thermokinetics to mechanics; on the contrary, it constitutes the broader framework within which mechanics has its own place. The reversible processes studied within mechanics are considered to be a special case of the processes studied within phenomenological thermokinetics.

Roughly, Stueckelberg proceeds in the following way. He assumes axiomatically that the state of a thermodynamical system is completely described by its 'geometrical', i.e. mechanical, variables and (at least) one non-geometrical variable. For the latter, Stueckelberg chooses the entropy S instead of, for instance, the absolute temperature T, because, contrary to T, S is an extensive variable and, in that respect, of the same nature as the energy. Because S is a non-geometrical variable, it cannot be written as a function of the mechanical variables, more particularly of the phase space variables x_i and p_i. This implies that, in general, the thermokinetic description cannot be reduced to the mechanical description. The entropy S characterises the 'internal' state of the system [op. cit. p. 28].

The evolution of a system is governed by the two famous principles of phenomenological thermodynamics, from which the equations of motion for all the state variables, both geometrical and non-geometrical, must be derived. To this end, the first principle concerning the conservation of energy, is generalised so as to encompass all conserved extensive quantities, including not only the conservation of energy but also the conservation of linear and angular momentum (and moreover, the conservation of mass for the non-relativistic case) [op. cit. p. 25 ff]. In other words, the new first principle of phenomenological thermodynamics embraces the mechanical conservation laws; this is Stueckelberg's first step toward a combined thermodynamical-mechanical description of physical reality. With regard to the second principle, the first principle has a very special function, for it specifies the constraints under which the maximalisation of entropy occurs. Given any arbitrary system, this maximalisation must take place so that quantities like energy, mass and momentum are conserved. Stueckelberg's reformulation of the second principle of phenomenological thermodynamics is especially relevant, since it leads to the conclusion that the origin of thermodynamic time asymmetry is to be found within this principle.

According to Stueckelberg, the second principle in its usual form,

$$dS/dt \geq 0,$$

cannot account for the irreversible behaviour of thermodynamical systems because the action of the time reversal operator T* on the entropy S can be defined consistently so as to render this law time reversal invariant (see p. 128). Therefore he proposes [op. cit. p. 23] to reformulate it in two sub-principles: 1) the principle of evolution, which is symmetric in time, and 2) the principle of equilibirum, which introduces a difference between past and future orientation of time. According to this reformulated principle, there exists for any system and extensive state function S, called 'entropy', which satisfies two conditions:

1) the principle of evolution: if the system is adiabatically closed, then S is a monotonic non-decreasing function of time,

$$dS/dt \geq 0,$$

2) the principle of equilibrium: if the (finite) system is isolated, then S tends toward a finite maximum in the distant future:

$$\lim_{t \to \infty} S \longrightarrow S_{max}$$

Note that the principle of evolution applies to adiabatically closed systems (i.e. they may exchange energy with their environment in the form of mechanical work, but not in the form of heat), whereas the principle of equilibrium is valid for isolated or totally closed systems.

The principle of evolution does not, for the reason given above, lead to any asymmetry in time. Instead, the principle of equilibrium breaks the symmetry between the past and future orientation of time. The reason lies in the fact that while entropy does have a maximum value, it does not have a minimum value. In order to see why this is the case, the so-called third principle of phenomenological thermodynamics has to be taken into consideration. It states that the entropy has a lower bound at the zero point of absolute temperature T; but a lower bound is not a minimum. Under time reversal, therefore, the maximum of entropy is not transferred into a maximum but into an upper bound. This destroys the symmetry between both orientations of time; entropy has a maximum relatively to one orientation of time only [op. cit. p. 24].

From this reformulated second principle, Stueckelberg derives the dissymmetry in the solutions of the equations of motion, viz., their convergence relatively to the future orientation of time and their divergence relatively to the past orientation. In a nutshell, he argues as follows. A direct consequence of the reformulated second principle is that it imposes certain restrictions upon the signs of quantities appearing in the equations of motion of a system. On the one hand, it requires that thermal coefficients, like heat capacity, have a positive sign regardless of the sign of absolute temperature T; on the other hand, it entails that mechanical quantities like mass and the constants of elasticity and friction have the same sign as absolute temperature [op. cit. p. 32-33, p. 43]. The determination of the signs of these quantities relative to the sign of absolute temperature leads to the dissymmetry in the solutions of the equations of motion and gives rise to the direction of time.

Note that the determination of the signs of constants like heat capacity and friction coefficient is not a matter of arbitrary convention, but, on the contrary, follows from the second principle. Therefore, the convergence of the solutions of the equations of motion toward the future, and their divergence toward the past, does not result from an appropriate choice of the signs of these quantities, but is a consequence of the second principle which thereby invalidates Laplacean determinism.

Stueckelberg illustrates the foregoing with the example of the one-dimensional harmonic oscillator with damping [op. cit. p. 33-36]. When such oscillator is considered to be a thermodynamical system, then its state is completely characterised by its entropy S which is the only non-geometrical variable, and by its geometrical variables r and v, representing respectively position and velocity of the mass M. Its energy H is given by:

$$H = H(S,r,v) = 1/2Mv^2 + U(S,r),$$

where U(S,r) represents the potential or 'internal' energy of the system. From the first principle, Stueckelberg infers that there must be a balance between the Newtonian inertial, the frictional (F^{fr}) and the elastic (F^{el}) force:

$$M dv/dt - F^{fr}(S,r,v) - F^{el}(S,r,v) = 0,$$

where F^{fr} is, by definition, equal to $-\lambda v$, with λ being the friction-coefficient. The principle of evolution implies that:

$$\frac{dS}{dt} = \frac{\lambda}{T}(S,r,v)v^2 \geq 0,$$

from which it follows that the friction coefficient λ must have the same sign as absolute temperature T. Similarly, the principle of equilibrium fixes the sign of the mass M and the constant of elasticity a (20). Next, Stueckelberg considers the behaviour of the oscillator in the neighbourhood of the equilibrium state which is given by ($S=S_{max}$, $r=r_0(S)$, $v=0$). In the linear approximation of this state, the energy can be written as:

$$H(S,r,v) = \tfrac{1}{2}Mv^2 + \tfrac{1}{2}a_0(r-r_0)^2 + U_0,$$

where a_0, r_0 and U_0 are functions of S only. This leads to the following differential equation for r:

$$\frac{d^2}{dt^2}(r-r_0) + 2\gamma\frac{d}{dt}(r-r_0) + \omega_0^2(r-r_0) = 0,$$

with

$$2\gamma = \frac{\lambda_0}{M} \geq 0 \quad \text{and} \quad \omega_0^2 = \frac{a_0}{M} \geq 0.$$

Whether the damping be strong ($\gamma^2-\omega_0^2>0$) or weak ($\gamma^2-\omega_0^2<0$) the solutions of this equation do indeed converge toward the distant future (i.e. for t goes to plus infinity). This convergence is not the result of an aposteriori selection of the appropriate signs for M, λ and a; instead, the determination of the signs of these quantities is based directly upon the second principle.

The foregoing shows that within Stueckelberg's approach the thermodynamical and mechanical description are closely intertwined (22). Reversible mechanical processes become a special case of irreversible thermodynamical processes; a purely mechanical description is only possible for those isolated systems for which the entropy is constant in time, in which case the process under consideration is reversible. On this point, Prigogine, also a defender of the thermodynamical program, strongly deviates from Stueckelberg's approach.

IV.3.2 Complementarity between mechanics and thermodynamics.

The central problem of Prigogine's book From being to becoming is: how can dynamics, and more particularly classical and quantum mechanics, 'the physics of being', be reconciled with phenomenological thermodynamics, the 'physics of becoming'. He tries to bridge the gap between these two by developing a 'microscopic theory of irreversible processes', from which dynamics and thermodynamics emerge as complementary modes of description. Within this theory, time no longer functions as mere parameter as it is the case in dynamics, but it now becomes an operator as well. In the following pages, I shall not go into the mathematical details of Prigogine's theory, but I shall concentrate upon its main ideas and upon his reasons for developing such theory. Fortunately, Prigogine extensively argues why the construction of a theory which embraces both dynamics and thermodynamics is necessary. This is of primary interest for the topic of this chapter.

Prigogine calls dynamics the physics of being because it eliminates 'becoming' from the description of physical reality. In a dynamical description, there is no difference between past and future. Any dynamical process that can take place in the 'forward' direction of time can also take place in the opposite direction, and, given the state of a dynamical system at any aribitrary moment, its whole future and past are in principle determined because of Laplacian determinism. One point of the trajectory of the system in phase space determines the entire trajectory (this is related to the analyticity of the solutions of the dynamical differential equations which, as will be seen shortly, no longer obtains for dynamical systems showing the so-called Poincaré catastrophe). Thus within dynamics, time really functions only as a parameter which itself is "unaffected by the transformation that it describes" [Prigogine 1980, p. 3]. In the dynamical picture of the physical world, there is no place for irreversibility as described by the second law of phenomenological thermodynamics, and thus, no place for becoming. It is precisely because thermodynamics puts the accent on irreversibility that Prigogine associates it with a physics of becoming.

Prigogine is not satisfied with the usual solution of the conflict between the physics of being and the physics of becoming, which lies in conceiving thermodynamic irreversibility as corresponding to some sort of approximation added to the reversible laws of dynamics, or in considering it to be an illusion due to certain limitations in human observation, such as a lack of microscopic information (see sect. IV.2.3). Prigogine's own point of view with regard to irreversibility is summarised in the following statement which is the main thesis of his book [op. cit. p. XIII]:

> First, irreversible processes are as real as reversible ones; they do not correspond to supplementary approximations that we of necessity superpose upon time-reversible laws.
> Second, irreversible processes play a fundamental constructive role in the physical world; they are at the basis of important coherent processes that appear with particular clarity on the biological level.
> Third, irreversibility is deeply rooted in dynamics. One may say that irreversibility starts where the basic concepts of classical or quantum mechanics (such as trajectories or wave functions) cease to be observables. Irreversibility corresponds not to some supplementary approximation introduced into the laws of dynamics but to an embedding of dynamics within a vaster formalism.

Let us first go deeper into the last point of Prigogine's thesis.

For Prigogine, the occurrence of irreversible processes is connected with the dynamical nature of the systems involved. As is well known, different kinds of dynamical systems can be distinguished. For instance, they can be classified as integrable or non-integrable, depending on whether or nor the Hamiltonian H can be written in a suitable form (viz., $H = H(J_1, \ldots, J_n)$, where the J_i are the action variables), and as ergodic or non-ergodic, depending upon whether or not the point representing the system in phase space covers the whole hyperplane corresponding to the energy of the system. Of particular interest for the problem of irreversibility, is the classification of dynamical systems, based on the evolution of a volume element of phase space. The flow of such volume element can assume different forms which are represented schematically in fig. 23; each flow pattern corresponds to a different category of dynamical systems of increasing complexity, viz., non-ergodic, ergodic but non-mixing, and mixing systems (23). Other categories have been introduced recently as well, such as the so-called K- and C-systems of which the behaviour in phase space is even more complicated.

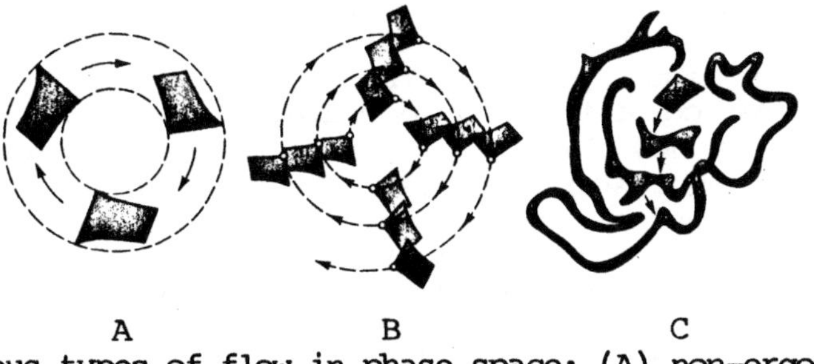

Various types of flow in phase space: (A) non-ergodic; (B) ergodic but non-mixing; (C) mixing.
fig. 23

(From From Being to Becoming by I. Prigogine. W.H. Freeman and Company. Copyright © 1980.)

The more complex a dynamical system is, the more unpredictible is its behaviour and the more its behaviour resembles the behaviour of stochastic processes. Features of a randomlike behaviour also appear in the case of dynamical systems which exhibit the so-called Poincaré catastrophe; in the Poincaré catastrophe the invariants of motion, except for the Hamiltonian itself, cannot be written as analytic functions of the action variables [op. cit. p. 41]. Thus, several types of dynamical systems, such as mixing and K-systems and those exhibiting the Poincaré catastrophe, show a randomlike behaviour. According to Prigogine, these kinds of dynamical systems are of special interest; for these systems "the dynamical motion introduces random elements" [op. cit. p.43].

A closer look at the trajectories of these complex dynamical systems in phase space reveals that two trajectories which are initially arbitrarily close together may separate strongly in the course of time. This means that two systems which are initially indistinguishable from a macroscopic point of view, may evolve after some time into two different macroscopic states. This feature of the flow of trajectories is called 'weak stability' [op. cit. p. 43 ff]. Weak stability has important consequences for the predictibility of the behaviour of dynamical systems. Since it is not possible to give the exact location of a system in phase space, a precise determination of the future behaviour of a weakly stable dynamical system cannot be achieved. It is not possible to predict with certainty the future state of such system by reducing the inaccuracy in the determination of the initial state; however small this inaccuracy may become, the system may always evolve to different macroscopic states. The behaviour of a weakly stable dynamical system cannot be computed in a deterministic way. According to Prigogine, this fact "forces us to transgress the limits of a purely dynamical description" [op. cit. p. 44]. In his opinion, weak stability allows for the first step toward the introduction of irreversibility in the formal structure of dynamics, because this weak stability is closely connected with 'randomness' at the microscopic level, which emerges as irreversibility on the macroscopic level.

Given the fact that almost all dynamical systems are weakly stable [op. cit. p. 208] and that the state of a system can only be determined with a finite accuracy, Prigogine concludes that the notion of trajectory, so fundamental for classical dynamics, looses its meaning. The use of ensemble theory becomes unavoidable and the state of a system can only be described in terms of a distribution function in phase space, ρ, representing the probability density [op. cit. p. 207-208]. The 'motion' of ρ is given by the Liouville equation:

$$\frac{\delta \rho}{\delta t} = -\sum_{i=1}^{s} \left(\frac{\delta H}{\delta p_i}\frac{\delta \rho}{\delta q_i} - \frac{\delta H}{\delta q_i}\frac{\delta \rho}{\delta p_i}\right) = -iL\rho$$

with H the Hamiltonian of the system and L the linear Liouville operator [op. cit. p. 25]:

$$L = -i\frac{\delta H}{\delta p_i}\frac{\delta}{\delta q_i} + i\frac{\delta H}{\delta q_i}\frac{\delta}{\delta p_i}$$

Because of weak stability it is not possible to go from the description in terms of distribution functions to the description in terms of trajectories; both modes of description are no longer equivalent.

The foregoing makes it necessary to reconsider the status of the recurrence theorem of Poincaré. This theorem is of fundamental importance for the problem of the relation between the physics of becoming (thermodynamics) and the physics of being (dynamics), since it implies that thermodynamics and dynamics are incompatible in the sense that an entropy function S of momenta and coordinates such that $dS/dt \geq 0$, cannot be defined (see p. 149). But this recurrence theorem is valid for a description of dynamical systems in terms of trajectories. Now, if the notion of trajectory becomes meaningless for weakly stable dynamical systems, then at first sight, the most obstinate stumbling block for combining thermodynamics with dynamics into a unique scheme disappears. Unfortunately, however, the recurrence theorem still holds for a description in terms of distribution functions: the entropy cannot be defined as an average over a distribution function [op. cit. p. 171ff].

How, then, can we arrive at a dynamical interpretation of the second law of phenomenological thermodynamics? In a nutshell, Prigogine proposes the following solution. Instead of defining entropy, like Boltzmann, as a function of the phase space variables x_i and p_i, he introduces an operator M for the microscopic entropy of a dynamical system. The choice of an operator for microscopic entropy is related to the fact that complex dynamical systems cannot be described in terms of trajectories. This operator M does not commute with the Liouville operator L,

$$-i(LM - ML) = D \leq 0,$$

where D is defined as 'the microscopic entropy production' [op. cit. p. 174]. Because M does not commute with L, a new kind of complementarity arises between the thermodynamical and the dynamical description [op. cit. p. 174-175]:

> Either we consider eigenfunctions of the Liouville operator to determine the dynamical evolution of the system or we consider eigenfunctions of M, but there are no common eigenfunctions of the two noncommuting operators L and M.

Either the entropy operator M can be defined, in which case the system under consideration is not purely mechanical and does not admit of a description in terms of trajectories, so that a recourse to distribution functions is necessary; or no such operator M exists, in which case the system can be described in a purely mechanical way by the use of trajectories. Prigogine claims that such operator M can indeed be constructed for weakly stable dynamical systems [op. cit. p. 176]. They are 'sufficiently complex' to show a thermodynamical type of behaviour.

The further details of Prigogine's construction of a microscopic theory of irreversibility do not concern us here. However, one specific aspect of this theory must be mentioned, viz., the introduction of time as an operator. Along with the operator M, Prigogine introduces the operator T which is defined in the following way (T is the 'square root' of M):

$$M = T^{+}T$$

with T^{+} the Hermitian conjugate of T. This operator T has an interesting property; its commutator with L is constant [op. cit. p. 188]:

$$[L,T] = iI,$$

with I the unity operator. Since the Liouville operator L corresponds to the time derivative, the operator T is most naturally interpreted as corresponding "to a 'time' in the sense that the representation $L \rightarrow i\delta/\delta t$, $T \rightarrow t$, satisfies the commutation relation" [ibidem]. This new time, corresponding to the operator T, is interpreted by

Prigogine as a 'fluctuating time'; it represents the 'internal' time or 'age' of a system, and is therefore not merely a parameter like time in classical mechanics. Since

$$\frac{d}{dt}\langle T\rangle = \text{constant},$$

with $\langle T\rangle$ the expectation value of T, macroscopic, parameter time is "simply the average over the new operator time" [op. cit. p. 209]. According to this new concept of operator time, the age of a system depends upon the distribution function by which that system is described and is therefore not just an index or an external parameter.

I shall now turn briefly to the second part of Prigogine's main thesis stating that irreversible processes play an important constructive role in our universe. This is a rather remarkable claim since the second law of phenomenological thermodynamics has generally been interpreted as leading to a maximalisation of disorder in our universe and as expressing a tendency toward uniformity in nature: it is supposed to lead to a destruction of highly ordered, complex systems or structures.

Prigogine points out that this standard interpretation of the second law is only valid within a particular domain of phenomenological thermodynamics, viz., the study of closed systems in the neighbourhood of equilibrium. For these systems indeed, the state of maximum entropy is the state of highest uniformity and disorder. To show that the second law can contribute to a better understanding of the genesis of complex structures in our universe, Prigogine turns to the study of open thermodynamical systems far removed from a state of equilibrium. His thesis is that the creation of order occurs in open systems far from equilibrium, in which case fluctuations play an important role ('order through fluctuations'), whereas the destruction of order occurs in closed systems in the neighbourhood of equilibrium ('Boltzmann order principle').

The following example illustrates that non-equilibrium may be a source of order [op. cit. p. 81]:

> When we apply a thermal gradient to a mixture of two different gases, we observe an increment of

> one of the components at the hot wall, whereas the other concentrates at the cold wall. This phenomenon, already observed in the nineteenth century, is called thermal diffusion. In the steady state, the entropy is generally lower than it would be in a uniform state. This shows that nonequilibrium may be a source of order.

This observation has been the starting point for Prigogine's ideas. For more complex systems, the non-equilibrium state may generate a new kind of structures, the so-called dissipative structures. Dissipative structures may arise in complex open systems far from equilibrium. They stabilise their situation by a constant exchange of matter and energy with their environment [op. cit. p. 90]. In this way, they are able to maintain the non-equilibrium with their environment. Their total entropy change can be split into two parts, namely the internal entropy production which according to the second law is always positive or zero, and the entropy flow from outside which can be negative or positive (see fig. 24). Thus for a system that is not in equilibrium with its environment, it is possible to keep the total entropy change equal to zero and, therefore, to maintain the non-equilibrium by compensating the internal entropy production by an outward flow of entropy. This, however, implies that the system must be able to exchange matter and energy with its environment.

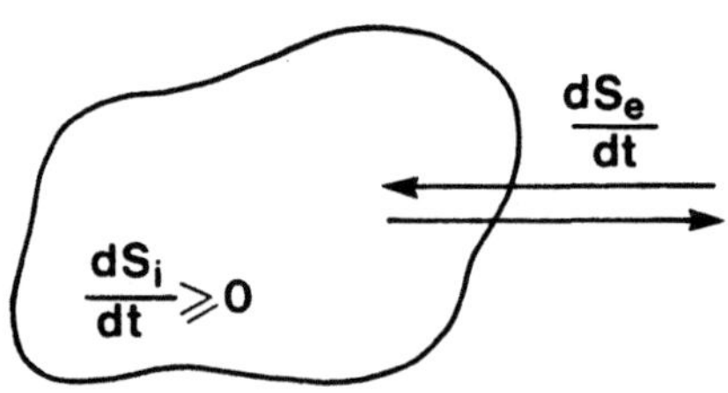

fig. 24

For Prigogine, these dissipative structures are "essential in the understanding of coherence and organisation in the nonequilibrium world in which we live" [op. cit. p. 84]. The study of the behaviour of dissipative structures and, more generally, of open thermodynamical systems far from equilibrium shows that their behaviour is very specific; it depends upon the particular constitution of the system involved. But this is precisely what is needed for an understanding of the abundance of different kinds of organisation in our world.

On the basis of complex chemical reaction schemes, Prigogine analyses the conditions under which dissipative structures may emerge; in his opinion, the occurrence of 'autocatalytic' reactions is a necessary condition. Since this type of reaction is frequently encountered in biological systems where they play an important role, Prigogine claims that the study of dissipative structures may contribute to a better insight into the structure of biological organisms.

Generally, the occurrence of instabilities in the behaviour of thermodynamical systems is of primary importance for the emergence of dissipative structures. Only when a system is sufficiently complex can such instabilities occur far from equilibrium. Starting from these considerations, Prigogine develops a theory of self-organisation mainly based upon stability theory [op. cit. p. 103ff]. One of the most striking aspects of this theory is the role which Prigogine attributes to fluctuations. Whereas generally, fluctuations play a minor role in macroscopic physics, they become crucially important in the theory of self-organisation. For large systems near equilibrium, the law of large numbers implies that fluctuations are neglegible. But in the neighbourhood of instabilities, the law of large numbers breaks down [op. cit. p. 131ff]. Whereas fluctuations near equilibrium are damped, non-equilibrium fluctuations in the neighbourhood of instabilities may become amplified and may determine which transition to a stable state will occur. Near equilibrium, fluctuations do not affect the average values of the relevant quantities, but in the neighbourhood of instabilities a "fluctuation drives the average" [op. cit. p. 132]. This leads to a new order principle [Prigogine 1973, p. 587]:

> In all these situations a <u>new order principle appears that corresponds essentially to an amplification of fluctuations and to their ultimate stabilization by the flow of matter and</u>

> energy from the surroundings. We may call this principle "order through fluctuations".

Thus, in the neighbourhood of instabilities, fluctuations determine the future evolution of the system. Once fluctuations have triggered off a transition to a stable state, the same fluctuations cannot bring about the reversed process, since fluctuations are no longer important as soon as the system has reached its final stable, equilibrium state.

To sum up, Prigogine claims that thermodynamics does not contradict the emergence of highly ordered systems in our universe. In order to explain the appearance of these systems not in equilibrium with their environment, it is not necessary to introduce "some new principle of nature, such as the 'instability of the homogeneous' or 'a differentiating force, creator of organization'" [Prigogine 1980, p. 84], nor to introduce an 'army of Maxwell demons' [Prigogine 1973, p. 589] that would have to account for the allegedly anti-thermodynamical behaviour. On the contrary, their emergence is in accordance with the laws of physics, and particularly with the laws of phenomenological thermodynamics, when they are applied far away from equilibrium conditions.

So far, the basic ideas developed by Stueckelberg and Prigogine within the context of the thermodynamical program. In the discussion of these ideas in the next section, I shall treat extensively of the first part of Prigogine's main thesis where he claims that "irreversible processes are as real as reversible ones".

IV.4 Discussion

The mechanical conception of irreversibility as an accidental feature of our universe is diametrically opposed to the thermodynamical conception according to which irreversibility is deeply anchored in the fundamental laws of physics. The controversy concerning the two programs crystallises around the status of the second law of phenomenological thermodynamics: is it or is it not one of the fundamental laws of physics, the equal of the laws of mechanics?

Compared with the mechanical program, the thermodynamical program has received little attention from both physicists and philosophers of science. Undoubtedly, one of the reasons thereof lies in the nature of the thermodynamical program itself: it presents a strong conceptual break with traditional physics and requires a 'rethinking' of its foundations.

An evaluation of both programs is not an easy matter. In the first place, the thermodynamical program has not been developed to the same degree of sophistication as the mechanical program. Up to this day, the need for working on and elaborating the thermodynamical program has not been felt by the majority of physicists. From the point of view of the history of physics, this is not surprising. The work spent on the mechanical program has had a very fruitful impact upon the development of physics [Prigogine 1980, p. 45]. The challenge of thermodynamical irreversibility led to statistical mechanics, more particularly, to ensemble theory and ergodic theory. This creative role is by itself already a strong justification for the widespread commitment to the mechanical program. This program has proven to be highly 'progressive' in the past.

In the second place, the controversy regarding both research programs is mainly of a conceptual nature. It is not based upon a dispute about empirical data, nor is it to be expected that a 'crucial experiment' will decide the issue. In fact, there is no disagreement about the irreversibility of the relevant processes and about the adequacy of the second law of phenomenological thermodynamics in describing this irreversible behaviour. There is disagreement, however, about the interpretation and the integration of this second law within the whole building of physical theories.

Therefore, the issue concerns primarily the coherence of the physicist's conception of thermodynamical time asymmetry. Thus one might be tempted to evaluate both programs with respect to their 'inner conceptual coherence'. I am rather sceptical about the prospects of such an undertaking. Apart from the fact that a fair comparison would require a further articulation of the thermodynamical program, there are no objective criteria for comparing the inner coherence of conceptual frameworks (except trivial ones like logical consistency). A conceptual scheme in which thermodynamics and mechanics are related to each other by way of a reduction - whether in the form of a reduction of thermodynamics to mechanics as in the mechanical program, or in the form of the incorporation of mechanics into thermodynamics, from which pure mechanics emerges as a special case, as in Stueckelberg's approach - could be as coherent as a scheme in which thermodynamics and mechanics correspond to complementary modes of description of physical reality, as in Prigogine's case. Physicists have learnt 'to live', albeit reluctantly, with complementarity, and, generally, they no longer consider complementarity as a deficiency of a theory.

Finally, a proper evaluation of both programs is not possible without taking into account the other types of time asymmetry (see p. 139 ff). The question whether or not irreversibility should be built into the fundamental laws of physics is not specific to thermodynamic time asymmetry. It also presents itself for the other types of time asymmetry. Therefore, the mechanical and thermodynamical program would have to be evaluated within a broader perspective.

At the present time, an appraisal of the mechanical and the thermodynamical program would be premature. But Prigogine has challenged the defenders of the mechanical program to a confrontation, not only by offering a theory of microscopic irreversibility, but also by arguing that the mechanical conception of thermodynamical irreversibility is not acceptable, since it reduces irreversibility to an illusion or an approximation. Later on, I shall comment on his criticism against the mechanical program. First I would like to make some general observations on both programs, which, I hope, will contribute to a clarification of the controversy with regard to them. These observations concern i) the distinction between time asymmetry and the flow of time, ii) the distinction between reversible and irreversible processes, iii) the role of the reversibility objections, in particular of the recurrence theorem, and iv) the role of boundary conditions.

i)In the first place, it should be kept in mind that the real issue at stake is the problem of time asymmetry, and not the problem of the flow of time. In Stueckelberg's theory, it is immediately clear that the principle of equilibrium (see p. 164) only breaks the symmetry between past and future orientation of time.

From this principle it follows that the solutions of the equations of motion of a system converge relatively to one orientation of time, but diverge relatively to the other orientation. Thus, the solutions of the differential equations are time asymmetric. But Stueckelberg goes yet one step further; he assumes that the future orientation of time coincides with the orientation relatively to which the solutions converge. In other words, equilibrium is reached in the future. In this way, Stueckelberg also determines the direction of the flow of time. From a purely physical point of view, however, this determination of the direction of the flow of time is arbitrary; physically, there is no reason why the future orientation of time could not be defined as the orientation relatively to which the solutions diverge. For Stueckelberg, the contention that equilibrium is reached in the future is founded in phenomenology which implies an appeal to the experience of the flow of time by a conscious observer.

A similar remark applies to Prigogine's analysis; his failure to distinguish between time asymmetry and the flow of time has seduced him into entitling his book 'From being to becoming' and into referring to thermodynamics as the physics of becoming, whereas in fact he discusses the problem of time asymmetry rather than the problem of the flow of time or of temporal becoming. It is not surprising, therefore, to see that Prigogine gets into trouble as soon as he starts discussing the relation between the flow of time and the entropy increase in our world [Prigogine 1980, p. 212-213]:

> We may construct a Lyapounov function [an entropy function] that increases monotonously with the "flow" of time or another one that decreases [....] in one description equilibrium is reached in the "future", and in the other in the "past". In other words, the time symmetry of dynamics can be broken in two ways; however, how to distinguish between them is a difficult question.

Indeed, thermodynamical time asymmetry does not determine whether the forward flow of time coincides with the direction in which entropy increases or whether it coincides with the direction in which entropy decreases. Thermodynamical time asymmetry structurally distinguishes between both orientations of time; it does not, however, allow us to tell the past from the future.

ii) One of the most conspicuous differences between the two programs concerns the distinction between reversible and irreversible processes. According to the mechanical point of view, irreversibility is ultimately a matter of probabilities: all processes are in principle (relatively to the fundamental laws of nature) reversible, but the occurrence of the reverse of certain processes is so unlikely that they never take place de facto. Because irreversibility is a matter of probabilities, there is no clear cut distinction between reversible and irreversible processes, but there is a gradual transition from one type of process to the other (see the discussion about the (un)ordered deck of cards at p. 119-120).

This is not the case for the thermodynamical conception of irreversibility: isolated systems behave irreversible, because their behaviour is governed by a time reversal non invariant law. Either the entropy is constant during a process, which implies that the process is reversible; or the entropy increases, which means that the process is irreversible. It might be argued that this sharp distinction between reversible and irreversible processes could be more refined by conceiving of the amount of entropy increase as a quantitative measure of irreversibility. This would allow for the introduction of 'degrees of irreversibility'. But this would be rather problematic. It is not clear under what conditions the increase of entropy is small, because there is no absolute unit of entropy increase. Moreover, S is an extensive quantity and, therefore, the entropy production in any kind of process can be increased by scaling up. In that case, the process under consideration would become 'more irreversible' due to a higher entropy increase, whereas qualitatively the process does not change. It is equally unclear how the introduction of degrees of irreversibility could be reconciled with the idea that irreversibility has a lawlike character. It seems that the conception of irreversibility of the thermodynamical program implies that, in principle, there is a strict boundary between reversible and irreversible processes.

In this context, it is worthwhile to point out an ambiguity in Prigogine's theory. On the one hand, there is clear evidence that he is in favour of a clearcut distinction between reversible and irreversible processes. For instance, he writes that if we give up the second law of phenomenological thermodynamics and replace it by the statistical mechanical H theorem, then we are lost, for one could no longer distinguish between reversible and irreversible processes [op. cit. p. 174]. Similarly, the complementarity between the mechanical and thermodynamical description leads to a clearcut distinction between reversible and irreversible processes: either a microscopic entropy operator exists, in which case the system shows an irreversible, thermodynamical behaviour, or it does not exist, in which case it does show a reversible, mechanical behaviour [op. cit. p. 196-197].

On the other hand, Prigogine's analysis of the limits of a mechanical description suggests that there is a smooth, gradual transition from a mechanical to a thermodynamical type of behaviour. In order to show a thermodynamical behaviour, a dynamical system must be 'sufficiently' complex; the more complex it is, the more it shows a randomlike behaviour which leads to irreversibility. This creates the impression that there is no definite turning point at which there is a qualitative break between a mechanical and thermodynamical description. Instead, it points to a smooth, gradual transition from one mode of description to the other. In that case, there can be no sharp distinction between reversible and irreversible processes.

iii) The role of the reversibility objections, particularly of the recurrence theorem of Poincaré, should not be overestimated or misunderstood in a comparison of the two programs. These reversibility objections have played a major part in the development of the mechanical program. But as soon as a strictly mechanical point of view is abandoned, as in the thermodynamical program, they loose all their strength. The reversibility objections cannot be leveled against Stueckelberg's treatment of a thermodynamical-mechanical system because he assumes from the outset that the state of such system is characterised by at least one non mechanical variable, namely the entropy, over and above its standard mechanical variables. Axiomatically, the entropy S cannot be written as a function of the mechanical state variables, and therefore the reversibility objections do not apply. Prigogine proceeds differently; realising that the recurrence theorem only applies to functions of the phase space variables (or of distribution functions), he proposes to define microscopic entropy by way of an operator. Whenever this operator can be defined, the mechanical description and, therefore, the reversibility objections lose their meaning.

iv) Within the mechanical program, the boundary conditions stated in the branch systems hypothesis perform a double function. In the first place, these boundary conditions are necessary in order to account for the irreversible behaviour toward equilibrium of low entropy systems. This is achieved by assuming that the formation of low entropy branch systems satisfies certain conditions of randomness [Grünbaum 1973, p. 256]; from these it follows that an arbitrarily selected low entropy branch system is most probably at the bottom of a dip in its entropy curve and that it therefore will approach equilibrium. In the second place, these boundary conditions posit the actual creation of low entropy branch systems in our part of the universe; in this way, the time asymmetric behaviour of thermodynamical systems as observed in daily life can be made plausible. The boundary conditions also assure the constant production of low entropy systems without which the actual occurrence of thermodynamical time asymmetry in our region of the universe cannot be understood.

Within the thermodynamical program, boundary conditions play a less prominent role. It is not necessary to introduce special assumptions about boundary conditions in order to explain the irreversible behaviour of thermodynamical systems. Given a system in a relatively low state of entropy, its irreversible behaviour is the result of the second law of phenomenological thermodynamics and does not require special restrictions upon the initial low entropy state. There is no intrinsic connection between the initial state of the system and its subsequent irreversible behaviour as in the mechanical program. Boundary conditions are important, however, with regard to the problem of explaining the actual occurrence of thermodynamical time asymmetry in our region of the universe. With respect to this problem, boundary conditions play the same role in both programs. Thus, the thermodynamical program also requires the introduction of some kind of branch systems hypothesis whose function, however, will not be the same as in the mechanical program. It is necessary only for an account of the actual occurrence of thermodynamical time asymmetry in our region of the universe, not for an explanation of the irreversible behaviour of low entropy systems.

The last three points illustrate once more that the thermodynamical and the mechanical program follow completely different paths in solving the problem of thermodynamical time asymmetry. With respect to traditional physics, the most interesting and challenging feature of the thermodynamical program is that it not only expresses the need for reconsidering the foundations of physics in connection with the problem of irreversibility, but that it also shows some ways in which the structure of mechanics could possibly be extended or embedded in a broader framework so that irreversibility is built into the foundations of physics. Whether or not an 'irreversible physics' can be reconstructed upon these new foundations so that it can compete with 'reversible physics', is at this moment an open question. Both Stueckelberg and Prigogine are convinced that it is worthwhile to proceed in this way in order to 'save' irreversibility. It seems that most physicists do not share this conviction and question the need for such extension of mechanics. They are convinced that irreversibility is ultimately a matter of probabilities and that therefore thermodynamic irreversibility can be accounted for in a purely mechanical way. But precisely on this point, they are criticised by the defenders of the thermodynamical program, who claim that irreversibility is thus degraded into an illusion. Especially Prigogine has attacked the mechanical conception of thermodynamical time asymmetry for essentially this reason. His arguments against the mechanical program will now be examined in more detail; I shall argue that they are not convincing and do not endanger the mechanical program.

In the first part of his main thesis (see p. 168), Prigogine states that "irreversible processes are as real as reversible ones" and that irreversibility does not correspond to some kind of "supplementary approximations that we of necessity superpose upon time-reversible laws". He assumes that the second law of phenomenological thermodynamics is strictly valid; which implies that it must be possible to define, in one way or another, some kind of deterministic H theorem à la

Boltzmann which is exactly valid and is not the result of approximations added to the laws of mechanics. The statistical H theorem is not acceptable for two reasons. In the first place, it gives up the monotonic increase of entropy. "But then", Prigogine says, "we are lost, for the distinction between reversible and irreversible processes would have to be replaced by some new one, which at present we cannot even formulate in a consistent way" [1980, p. 174]. In the second place, the statistical H theorem is based upon techniques like coarse graining. But apart from the fact that this approximation technique has proven to be unsuccessful, it makes irreversibility into an anthropomorphic or subjectivistic phenomenon which would be due to 'poor' human observation [op. cit. p. 12, 157].

For Prigogine, the recurrence theorem implies that within a mechanical description a foundation to the second law of phenomenological thermodynamics cannot be given; within such description, thermodynamical irreversibility necessarily becomes an illusion or an approximation. Prigogine cannot accept this conception of irreversibility, and therefore, he regards the mechanical description of physical reality as inadequate because it does no justice to one of the principal features of the physical world, namely irreversibility. This is the main motive behind his attempt to develop a theory which trancends the limits of a purely mechanical description.

In my opinion, this criticism of the mechanical program is not justified. For Prigogine, a description of physical reality, in which irreversibility is not anchored in one of the fundamental laws, degrades irreversibility into an illusion or approximation. Irreversibility is 'real' only if it has a lawlike nature. However, a comparison of the statistical H theorem with the second law of phenomenological thermodynamics shows that this interpretation of what is 'real' is unnecessarily restrictive.

Clearly, the statistical H theorem is not completely equivalent to the second law; it is a kind of approximation of this law in the sense that, instead of predicting with absolute certainty that the entropy of a system not in equilibrium will increase or remains equal, it merely asserts that this will most probably be the case. But for non equilibrium macroscopic systems, this probability becomes practically equal to one because, once these systems have reached equilibrium, large fluctuations away from equilibrium are extremely improbable. Therefore, the process will appear to be de facto irreversible as is illustrated in the following quotation of Boltzmann's Lectures on gas theory [1964, p. 444]:

> One should not however imagine that two gases in a 1/10 liter container, initially unmixed, will mix, then again after a few days separate, then mix again, and so forth. On the contrary, one finds by the same principles which I used for a similar calculation that not until after a time enormously long compared to $10^{10^{10}}$ years will there be any noticeable unmixing of the gases. One may recognize that this is practically equivalent to never, if one recalls that in this length of time, according to the laws of probability, there will have been many years in which every inhabitant of a large country committed suicide, purely by accident, on the same day, or every building burned down at the same time - yet the insurance companies get along quite well by ignoring the possibility of such events. If a much smaller probability than this is not practically equivalent to impossibility, then no one can be sure that today will be followed by a night and then a day.

Thus, for macroscopic systems not in equilibrium, the predictions of the statistical H theorem are perfectly in accord with our observations; the statistical H theorem "is all that is needed to successfully account for the observations of the real world" [Davies 1977, p. 5]. For macroscopic systems there is in practice no difference between the predictions made on the basis of the statistical H theorem and those made on the basis of the second law of phenomenological thermodynamics. The latter in principle rules out the reversal of certain processes, whereas the former predicts that the reversal will never de facto occur because it is extremely improbable. In either case, the predicted behaviour will be the same, namely an irreversible approach to equilibrium. But then there is no reason at all to qualify the resulting irreversible behaviour as 'real' in one case, and as an 'illusion' in the other.

Of course, the statistical H theorem is derived on the basis of time reversal invariant laws, and therefore all processes are in principle (i.e. relatively to those laws) reversible. For Prigogine, this means that "we are lost" because one could no longer distinguish meaningfully between reversible and irreversible processes. But if we accept that irreversibility is of a de facto nature and give up the idea of a clear cut distinction between reversible and irreversible processes, then the situation does not look that hopeless, since after all it is possible to indicate under what conditions systems will show a de facto irreversible behaviour. As is well known, for gases this strongly depends upon the size of the system (the number of molecules involved) as is clearly illustrated by the following example given by Smoluchowski (24). In air at 300 $^{\circ}K$ with a density of 3.10^{19} molecules/cm^3, the average recurrence time for deviations of 1% of the average number of molecules in a sphere of radius 5.10^{-5} cm, is of the order of 3.10^{60} years. However, when the radius of the sphere is reduced to 1.10^{-5} cm, the average recurrence time for deviations of 1% becomes about 10^{-11} sec. Thus for 'large' macroscopic systems the probability for the spontaneous occurrence of significant deviations from the equilibrium state becomes so extremely small that de facto they will never take place. Consequently, once such macroscopic system has reached equilibrium, it will not spontaneously return to its initial state, and so, the reversal of the process by which it reached equilibrium will in fact not occur; it is improbable to the degree that it is "practically equivalent to impossibility".

Similarly, Prigogine's objections against the use of techniques like coarse graining are not convincing. According to him, the use of these techniques implies that irreversibility becomes an illusion or an approximation. But on closer inspection, this conclusion appears to be unfounded. It is based upon the assumption that in the statistical mechanical description, irreversibility is related to the introduction of coarse graining. This is not the case, for although the statistical H theorem is derived by coarse graining phase space, it nevertheless is completely symmetric in time (see p. 150 - 153). Therefore, a statistical mechanical description of a low entropy system predicts with or without coarse graining, a de facto irreversible behaviour [Davies 1977, p. 77-78]:

> It must not be imagined that to remove coarse graining removes the macroscopic asymmetry. A complete description of the pendulum motion would still predict, with overwhelming probability, that the pendulum would be damped to rest..

Within statistical mechanics, coarse graining has a special function: it is introduced in order to define a 'quantitative measure of irreversibility', i.e. to define an entropy function; but it is not the origin of the de facto irreversible behaviour of systems.

In this respect, it is important to recognise that a distinction must be made between the problem of explaining why certain systems behave in an irreversible way and the problem of defining a quantitative measure of irreversibility, like an entropy function S. Prigogine is very much concerned with the second problem; he is looking for an universally valid expression for the entropy of a system so that the second law of phenomenological thermodynamics is strictly valid. Since within a mechanical description such entropy function of the phase space variables x_i and p_i does not exists, he concludes that mechanics cannot account objectively for irreversibility. This conclusion does not follow. Even without introducing a quantitative measure of irreversibility, mechanics predicts that a closed, non equilibrium, macroscopic system will show a de facto irreversible behaviour toward equilibrium. The problem of defining a quantitative measure of irreversibility constitutes the next step. If it be assumed that an objective foundation of coarse graining can be given, then the coarse grained entropy function of statistical mechanics is likely to be a good candidate for a quantitative measure of irreversibility. According to Prigogine, however, coarse graining cannot be given an objective foundation. Assuming for the sake of argument that he is right, this means that the advocates of the mechanical program should try to find a different way of defining a quantitative measure of irreversibility; it does not mean that this program cannot account for the de facto irreversible behaviour of macroscopic systems.

These considerations show clearly that, from the mechanical point of view, there is no need to insist with Prigogine upon the strict validity of the second law of phenomenological thermodynamics. Such insistence endangers mechanics without yielding any profit, since the statistical H theorem leads to the same predictions with regard to the behaviour of non equilibrium systems. Moreover, it is not necessary to follow Prigogine's claim that weak stability forces us to "transgress the limits of a purely dynamical description" because, in his opinion, weak stability would make the notion of a trajectory meaningless. A description in terms of distribution functions, be they needed for practical reasons or in principle, can perfectly well be combined with the principles of mechanics, as is standard practice in statistical mechanics. The use of distribution functions does not by itself undermine the mechanical mode of description, even if in principle a transition to a description in terms of trajectories is out of the question.

In the foregoing pages, I have tried to show that the mechanical program can be defended successfully against the charges brought up against it from the side of the thermodynamical program. This conclusion should not be misunderstood. In the first place, it does not make the theories developed within the thermodynamical program obsolete. But further work on this program should not be justified primarily by arguments against the mechanical program; it must derive its rationale from its inner coherence especially with regard to the problem of thermodynamic time asymmetry, and it must show that it can serve as a fruitful point of departure for physics. "Who sees the future?", Boltzmann wrote in the introduction to his <u>Lectures on gas theory</u>, and he immediately added, "Let us have free scope for all directions of research; away with all dogmatism...." [1964, p. 26]. Moreover, a proliferation of theories regarding the origin of thermodynamic time asymmetry should also be welcomed by the defenders of the mechanical program, for precisely by competing with other theories can it prove its vigour.

Secondly, it does not mean that the explanation of the occurrence of thermodynamical time asymmetry in our region of the universe offered by the mechanical program is itself without problems. As I intend to show in the remaining pages, the mechanical solution of the problem of thermodynamical time asymmetry, in particular Reichenbach's branch systems approach, is in various ways incomplete.

We have seen that, given a system not in equilibrium, the mechanical program can explain why this system will almost certainly evolve toward equilibrium. Of course, the next question is: How, when we observe a low entropy system in daily life, did the system reach the state of low entropy in the first place? In other words, how can we explain the constant creation of low entropy systems in our surroundings? They cannot be the result of large, spontaneous fluctuations such as occur in closed systems. Indeed, such explanation would ruin the argument by which the irreversible behaviour toward equilibrium was deduced; non equilibrium systems irreversibly evolve toward equilibrium because large, spontaneous fluctuations away from equilibrium de facto do not take place. From the point of view of the behaviour of closed systems, however, transitions from non equilibrium to equilibrium are as rare as transitions in the reversed order, from equilibrium to non equilibrium. Thus, the really interesting question about processes showing thermodynamical time asymmetry is not why the reversal does not take place, but why they themselves do take place at all.

Reichenbach's branch systems hypothesis is supposed to solve this problem of the constant creation of low entropy systems in our region of the universe. But for several reasons, it appears premature to consider this branch systems approach as the final solution of the problem of thermodynamical time asymmetry.

To begin with, the first part of the branch systems hypothesis, in which the universe in our region is said to be in a relatively low state of entropy, is not yet well understood. There is still much disagreement about the cosmological origin of this relatively low state of entropy [Davies 1977, Ch. 7]. The status of this boundary condition becomes especially problematic when the gravitational interaction is taken into consideration, for in that case, any region in our universe containing a sufficient amount of matter would be in a relatively low state of entropy, since in view of the phenomenon of gravitational collapse a self-gravitating system has no natural equilibrium state. Consequently, it seems that this condition can always trivially be satisfied by considering a sufficiently large part of the universe. A further clarification of these points regarding the first part of the branch systems hypothesis is needed.

Secondly, Reichenbach assumes that the creation of low entropy systems in our surroundings is a well established empirical fact [1971, p. 117]:

> Our environment is rich in processes which, either as a natural product or through the intervention of man, create as parts of their results ordered subsystems, which from then on remain isolated and run through an evolution toward disorder...

Indeed, one cannot but agree with this. But strangely enough, Reichenbach does not take the next step; he makes no effort at all to make this constant creation of low entropy branch systems plausible (except for the phrase "either as a natural product or through the intervention of man"). He seems to be satisfied with having shown that the basic assumptions from which he derives the time asymmetric behaviour of the majority of branch systems, are 'factual statements' about the universe. But precisely these factual statements themselves require further explanation. The boundary conditions specified in the branch systems hypothesis are very special; they do not as usual merely describe some brute, contingent fact about nature, which defies any explanation, but instead, they state that nature displays a certain pattern by constantly creating low entropy branch systems. This pattern itself needs further clarification and explanation.

Finally, we must ask how it is possible to reconcile the irreversible behaviour of thermodynamical systems toward disorder (equilibrium) with the evolution from the presumed homogeneous and undifferentiated conditions in the early stages of the universe to the present, highly structured and differentiated state? This evolution to a higher state of organisation has taken place at different levels; there has been an evolution of the universe at the microscopic level, e.g. the creation of the chemical elements, but also at the macroscopic level, e.g. the organisation of the 'primordial soup' into solar systems, star clusters, galaxies, clusters of galaxies etc. This tendency toward ever more complex systems is particularly obvious in the biological evolution. From the point of view of the time asymmetric behaviour of thermodynamical systems toward disorder, this constant evolution toward ever more complex and highly structured systems is paradoxical.

Up to this date, the mechanical program has not provided an explanation of this extraordinary kind of time asymmetry. In this respect, the prospects for the thermodynamical program are more promising. According to Prigogine's theory of self-organisation, the creation of coherent structures does not contradict the laws of phenomenological thermodynamics; whereas the second law leads to a destruction of order in the neighbourhood of equilibrium, it may lead to a creation of order in open systems far from equilibrium. Thus, the emergence of order in our universe does not contradict the second law, but on the contrary, it is a consequence thereof.

For the advocates of he mechanical program it is important that they see that Prigogine's theory of self-organisation does not depend upon his theory of microscopic irreversibility. The theory of self-organisation is based on a phenomenolgical description of physical reality, more specifically, on phenomenological thermodynamics [Prigogine 1980, p. 77]. This means that a commitment to Prigogine's theory of self-organisation does not imply a commitment to his theory of microscopic irreversibility. Thus, the way is open for incorporating the theory of self-organisation into the mechanical program within which phenomenological thermodynamics is accepted as an adequate theory for describing the irreversible behaviour of macroscopic systems. It may be along this path that a coherent interpretation of the tendency toward both order and disorder in our universe can be achieved.

NOTES

1 Parts of this chapter have already appeared in Kroes [1982].

2 For more details, see for instance Davies [1977, Ch. 6] and Jammer [1974, Ch 11].

3 The implications of various cosmological models for the connections between different types of time asymmetry are extensively discussed in Davies [1977]. See also Y. Ne'eman, "Time reversal asymmetry at the fundamental level - and its reflection on the problem of the arrow of time" in Landsberg [1982].

4 At least in the standard conception of the operator T*; see p. 128.

5 Within a macroscopic or phenomenological description it is assumed that fluctuation phenomena are not significant and can be neglected. In general this is only possible if the systems under consideration are sufficiently large; in that case, quantities like temperature and pressure can be meaningfully defined in terms of definite experimental procedures.

6 In the following, I shall consider the relation between the second law of phenomenological thermodynamics and classical mechanics. The problem does not change substantially when classical mechanics is substituted by relativistic mechanics or quantum mechanics. Their significant common feature is that their laws or principles are all time reversal invariant.

7 Notions like 'research program' and 'hard core' are taken from Lakatos [1978]; roughly, a research program consists of i) a hard core of theories assumed to be irrefutable, and ii) a positive heuristic which defines the relevant problems and shows the way for constructing auxiliary hypotheses to protect the hard core. A research program is said to be progressive as long as it successfully predicts new facts and solves theoretical problems, but degenerative as soon as it produces only 'post hoc' or 'ad hoc' explanations. In the following I shall use these notions in a rather loose sense; it is not my intention to produce here a 'rational reconstruction' of the historical development of the problem of thermodynamic time asymmetry.

8 For more details on the topics discussed in this and the next section, see Davies [1977].

9 It is possible to show that, for infinite systems, a certain increase in entropy will occur by taking the so-called 'thermodynamic limit'; in that case, the statistical entropy goes over into the thermodynamic entropy [Cfr. Landsberg 1970].

10 For more details, see Davies [1977, p. 56] and the references given there.

11 For a derivation of the statistical H-theorem and the role of the ergodic hypothesis, see Davies [1977, p. 63 ff].

12 In most discussions of thermodynamic time asymmetry, the Boltzmann entropy plays a dominant role. It should, however, be noted that beside this Boltzmann entropy, other entropy concepts have been introduced, for instance, the coarse grained Gibbsean entropy. The latter type of entropy differs strongly from the Boltzmann entropy because its value does not depend upon the actual behaviour of a gas as for the Boltzmann entropy, but only depends upon the constraints imposed upon the gas [Sklar 1977, p. 393]. In the following, I shall refer to the Boltzmann entropy of a gas simply as the entropy of a gas.

13 The reason is that smaller deviations from equilibrium occur much more frequently than larger fluctuations [Davies 1977, p. 70, Grünbaum 1973, p. 257-258, Sklar 1977, p. 387, Scott 1967, p. 28 and the references given there].

14 For macroscopic systems T is of the order of $10^{10^{23}}$ sec. [Davies 1977, p. 56].

15 Grünbaum [1973, Ch. 19] has extensively discussed this problem.

16 Although, generally speaking, random outside perturbations are thus not relevant for the problem of thermodynamic time asymmetry, it wil be seen later on, sect. IV. 3 , that according to Prigogine, random disturbancies from outside or fluctuations occurring inside a system, under special conditions, do play a dominant role in the irreversible behaviour of certain thermodynamical systems.

17 Note that with respect to the statistical time asymmetric behaviour of ensembles of branch systems, it is important that the branch systems exist only for a short span of time (short relatively to the Poincaré

recurrence time of the systems). If the lifetime of these systems would be of the order of the recurrence time, then the effect of their initial low entropy states would disappear due to the subsequent, spontaneous occurrence of large deviations from equilibrium; in that case, the behaviour of an ensemble of branch systems would again be symmetric in time; cfr. Reichenbach [1971, p. 121] (inequality (6) holds only for small i, in other words, for small time intervals) and [op. cit. p. 126-127]; cfr. also Davies [1977, p. 78].

18 For a more detailed formulation of these boundary conditions, see Reichenbach [1971, p. 136].

19 The basic ideas underlying Stueckelberg's approach are not always explicitly stated in his book; for many details of the following exposition, I am endebted to many discussions I was privileged to have with P.B Scheurer who was co-author of Stueckelberg's book.

20 See also Weyl [1963, p. 205-206].

21 The condition of maximum entropy implies that for a system in equilibrium the first variation of entropy must be zero and the second variation negative. From these conditions, it can be derived that mass and the constant of elasticity must have the same sign as absolute temperature [op. cit. p. 32-35].

22 Stueckelberg also shows that the maximalisation of entropy under the constraints given by the first principle implies a minimalisation of mechanical energy and that the metric of space has to be definite positive [op. cit. p. 33, 106].

23 The Liouville operators corresponding to these various categories have different spectral properties [op. cit. p. 37, 188].

24 See Smoluchowski [1915] but also the discussion on p. 119 - 120.

EPILOGUE

OBJECTIVE VERSUS MINDDEPENDENT THEORIES OF TIME FLOW

In the study of time, the problem of the flow of time has a longstanding history (1). In particular, two issues with regard to the flow of time have been raised again and again:

i) whether there can be (a flow of) time without any change or becoming, and

ii) whether the flow of time and the corresponding becoming of events is real or a mere illusion. The first issue goes back as far as Aristotle who defined time "as the number of motion with respect to earlier and later" (2). Without any motion or change of some kind, there could be no time. Via e.g. Augustine, this problem has stayed alive up to the present day [Shoemaker 1969].

The second issue, to which this epilogue is devoted, can be traced back to at least Parmenides. He maintained that Being is One and "neither arises out of being nor out of not-being: it never came into being, but simply is" [Copleston 1962, p.67]. His reasoning led to the conclusion that change, becoming or motion are impossible; becoming and the flow of time are therefore mere illusions. Becoming is not real. This Parmenidean view re-emerged in our century within a scientific context rather than a philosophical context. More in particular, relativity theory has led to a revival of Parmenides' ideas because this theory, formulated within the four-dimensional spacetime formalism, has been interpreted in terms of a static four-dimensional block-universe in which there is no room for change or becoming (see sect. II.8). As a result of this, the second issue with regard to the flow of time has again received considerable attention.

In its modern setting, the dispute concerns the question whether the flow of time is an objective phenomenon or whether it is minddependent. Recently, both objective as well as minddependent theories of time flow (and becoming) have been put forward [Reichenbach 1956, McCall 1976, Grünbaum 1971]. The controversy about the flow of time concerns the question whether it is possible to introduce, on the basis of modern theories of physics, the notion of a flow of time in the physicist's conception of nature.

According to the objective theories of time flow, becoming is an aspect of physical reality and therefore the flow of time is a legitimate physical concept which in principle can be explicated without making an appeal to 'extra-physical' concepts such as 'consciousness' or 'mind'. This means that it must be possible to describe within a physical discourse an intrinsic difference between past events, present events and future events. In other words, the flow of time is an aspect of physical time and not just of psychological time.

The minddependent theories of time flow, on the other hand, maintain that nothing in physical reality corresponds to our experience of a flow of time and of becoming, and that talk about flow of time and becoming only makes sense within the context of the experience of a conscious observer. Ascribing becoming to the events themselves is as meaningless as the ascription of space and time to the 'Ding an sich' would have been within the Kantian system. Concepts such as 'past', 'present', 'future' and 'flow of time' are considered not to be genuine physical concepts. According to this point of view, the proper domain for studying these concepts is psychology, because they pertain to psychological time, and not to physical time.

In order to locate exactly the point of controversy between the objective and minddependent theories of time flow, I shall now present, in the order of increasing complexity, a brief outline of four notions of time, viz., mechanical time, evolutive time, historical time and individual time. It appears that from a physical point of view, mechanical, evolutive and historical time are not very problematic; it is generally assumed that they can be given a physical foundation, although there is still some disagreement about the manner in which this ought to be done. The controversy concerns the status of individual time within physics (3).

- Mechanical time. This notion stands for time as used by physicists in their theories of mechanics. Introduced by way of a variable t, mechanical time is usually considered to be isomorphic to the Euclidean line E. The most distinctive feature of mechanical time is that there is no difference between its past orientation (t-) and its future orientation (t+); t+ is mechanically equivalent to t-. The reason for this is that the laws of mechanics are time reversal invariant, and thus all mechanical processes are in principle reversible. Therefore it is impossible to distinguish between past and future orientation of mechanical time on the basis of the behaviour of mechanical systems.
This characteristic feature of mechanical time is closely connected with the circumstance that a mechanical description is 'acausal' in the sense that there is no real, intrinsic difference between 'cause' and 'effect'. Given an arbitrary mechanical process, it is not possible to rule out the time reversal of that process on causal grounds; the time reversed process does not violate the principle of causality. From the point of view of causality, t^+ is equivalent in mechanics to t^-. The cause-effect relation is asymmetric in time - the cause precedes the effect - and its presence in a mechanical description would therefore break the symmetry between both orientations of mechanical time.
- Evolutive time. Time will be called evolutive if i) there exists for every closed system some quantity A, the "parameter of evolution", which increases monotonically in time, or stays equal:

$$dA/dt \geq 0.$$

and ii) there is nevertheless no way of distinguishing between past and future orientation of time.

Time in mechanics does not satisfy the requirements of evolutive time. It is impossible to define, without violating the principles of mechanics, a mechanical parameter of evolution A, such that for all mechanical processes A increases (or stays equal) in time. In mechanical processes, therefore, no "real" evolution takes place. Within a phenomenological thermodynamical description, however, such a quantity, the entropy S, does exist. Given Stueckelberg's interpretation of the second law of phenomenological thermodynamics (see p.

164: the principle of evolution), the above requirements for evolutive time are satisfied. Note that in its standard interpretation, the second law does not satisfy the requirements for evolutive time, since in that case it is possible to distinguish unambiguously between both orientations of time.
As regards causality, the same applies to evolutive time as to mechanical time; no distinction between cause and effect is possible, since past and future orientation of evolutive time are equivalent.
- Historical time. Time will be called historical if it is possible to distinguish unambiguously between both orientations of time. Then t^+ and t^- are not physically equivalent, and one of them can be chosen as the 'future' orientation of time. Once this choice is made, it is possible to distinguish between the past (the history) and the future with regard to any arbitrary moment. But still all moments of historical time are equivalent insofar as the flow of time is concerned; no one is singled out as the present or the now.
It is clear that the existence of time asymmetry allows for an unambiguous determination of the past and future orientation of time since irreversible processes take place only relatively to one orientation of time. Thus, if a physical foundation can be given to irreversibility (or time asymmetry), then the distinction between past and future orientation of time also acquires a physical status. By virtue of the fixation of the past and future orientation, the distinction between cause and effect also acquires a meaning within physics. Within phenomenological thermodynamics, the emergence of causality manifests itself rather strikingly. One of the specific features of the cause-effect relation is that of deterministic connection: each cause necessarily leads to its effect which is uniquely determined by the cause. In a mechanical description there is a one-to-one correspondence between the initial and final states of a system: to every initial (final) state of the system there corresponds exactly one final (initial) state. Therefore, both the initial and final states of the system can be interpreted as causes without violating the requirement that each cause uniquely determines its effect. This is not true within a thermodynamical description. Consider two similar systems which are initially in different non-equilibrium states but which evolve to-

ward the same final state of equilibrium. Here we encounter a situation in which different initial states lead to the same final state (i.e. two different causes have the same effect). Under time reversal this would lead to the conclusion that one cause (the equilibrium state) has two different effects (two different initial states). Thus the deterministic character of causality would be violated. Within a phenomenological thermodynamical description, causality obtains with regard to one orientation of time.

- Individual time. The distinction between past and future orientation of historical time does not yet complete the structure of time. The final step toward the full notion of time is made by introducing the notions of past, present and future. All moments of historical time qualify as a potential present. In the experience of an individual, the actualisation of one of these as the present, with the corresponding past and future, takes place. But the actual present is continually changing and therefore individual time 'flows'. This cannot be said of mechanical, evolutive and historical time.

Thus, becoming makes sense only within the context of individual time with its distinction between past, present and future. The problem with regard to individual time is whether this aspect of becoming can be given a physical foundation; in other words, whether the distinction between past, present and future can be formulated within a physical discourse in the same way as it was possible for the distinction between past and future orientation of time. If so, then becoming would be an aspect of physical reality. The aim of the objective theories of time flow is to provide a physical foundation for the distinction between past, present and future.

Proponents of an objective time flow face the task of describing within purely physical terms an intrinsic difference between events belonging to the past, the present and the future. The difference between past, present and future events is commonly expressed in the following beliefs:

1) past events are fixed once and for all and are unalterable, whereas future events are not determined and can be influenced,

2) some of the past events can be remembered, but none of the future,

3) past events have been, present events exist and future events have still to come.

Now, none of these points can be reformulated into a physical discourse so as to allow a physically significant distinction between past, present and future (4). So a solution has to be found in another direction.

According to Reichenbach, quantum mechanics offers a way out. Stressing the indeterministic character of quantum mechanics, Reichenbach writes [1971, p. 269]:

> The concept of <u>becoming</u> acquires a meaning within physics: the present, which separates the future from the past, is the moment when that which was undetermined becomes determined, and 'becoming' means the same as 'becoming determined'.

In this way, Reichenbach tries to define a unique moment as the present without making an appeal to a conscious observer. Within the framework of classical Newtonian physics, a similar definition of the present is not possible because of Laplacean determinism: the present state of the universe determines the state of the universe at all other moments, in the past as well as in the future. But within the standard quantum mechanical description, the present state of the universe does not determine the future state of the universe. The future is still open. Whereas in principle it is possible to have records of the determined past events, this is not possible for the as yet undetermined future events. For Reichenbach, quantum mechanics gives an objective foundation to the fact that past events are determined and that future events are undetermined [Reichenbach 1971, p. 270].

Grünbaum [1971, 1973, ch. 10], however, has convincingly shown that Reichenbach's proposal is not acceptable. Given any arbitrary event E, events lying in the past of E are determined in the sense that records of them could exist at E, whereas events lying in the future of E are quantum mechanically undetermined relatively to E.

Thus it follows that [Grünbaum 1971, p. 221-222]:

> every event, be it that of Plato's birth or the birth of a person born in 2000 A.D., _at all times_ constitutes a divide in Reichenbach's sense between its own recordable past and its unpredictable future, _thereby satisfying Reichenbach's definition of 'the present' at any and all times!_

Thus Reichenbach's conception of objective time flow is inadequate for it does not single out a unique moment as the present.

A more recent attempt at constructing an objective theory of time flow has been undertaken by McCall [1976]. He claims that his theory is objective in the sense that "time flow characterises the universe independently of the existence of human beings". As for Reichenbach, the indeterminism of the future is of crucial importance in McCall's proposal. In case the universe were deterministic, his objective theory of time flow would break down.

McCall's theory is based upon the notion of a universe-picture which is a complete state description of the world including past, present and future. Each universe-picture has the form of a tree (without any roots): the stem represents the unique past, the branching point the present and the branched structure the undetermined future, with each branch corresponding to a possible course the future might take. Because the future is not yet determined relatively to the present, none of the branches of the future is in a preferred position. Different universe trees correspond to different moments of time (different presents); there are many "universe-pictures, each one being a complete state-description of the universe _at a time_." [McCall 1976, p. 342].

According to McCall, the description of the physical world in terms of universe-pictures has a dynamic character because the mode of branching of the universe changes in time [p. 343]:

> The dynamic feature of [this theory] ....consists in the following. The complete state-description of the universe, i.e., the

> universe-tree, is different at different times. The difference consists in this. If t2 is later than t1, the universe-tree of t2 is a proper subtree of the universe-tree at t1.

This dynamic feature which is to represent the flow of time, is objective for it does not depend, McCall claims, upon the presence of a conscious observer; rather this dynamic feature belongs to the ontological structure of the universe [ibidem]:

> Each moment of time, therefore, defines a separation of the universe into past and future that is ontological rather than epistemological. Furthermore, each moment in time defines a <u>different</u> separation. It is this feature ... that models time flow.

In this way, McCall tries to incorporate the flow of time into the description of physical reality. He also describes how his theory can be adapted so as to give a viable account of 'relativistic time flow'; within the present context, however the details of these modifications are not immediately relevant.

Instead, let us take a closer look at the alleged dynamic character of McCall's theory. For McCall, this dynamic character resides in the relation 'being a proper subtree of'. If t1 is later than t2, then the corresponding universe-tree of t1 is a proper subtree of the one of t2. But it is easy to demonstrate that the representation of the history of the universe by an ordered array of different universe-trees cannot give an adequate, dynamic model for the flow of time. The relation 'being a proper subtree of' does not convey a dynamic feature to the description of the physical world in terms of universe-trees for structurally this relation on the set of universe-trees is isomorphic to the 'smaller than' relation on the reals or the 'earlier than' relation on the moments of time: it is a total ordering relation. But an order relation as such can hardly be called a dynamic relation. If so, the temporal order relation 'earlier than' itself would be dynamic, and could therefore furnish a sufficient basis for describing the flow of time. There would be no need to locate the origin of the flow of time somewhere else.

But McCall's theory also fails with regard to the requirement of objectivity. As stated above, any objective theory of time flow has to meet the criterion that it singles out on physical grounds and in an unique way one moment of time as the present or the now. For this reason, Reichenbach's proposal turned out to be not acceptable. McCall is convinced that his theory is insensitive to the arguments of Grünbaum against Reichenbach for "the choice of which moment constitutes the present is not arbitrary but forced" [p. 346-347]. Of course, this is true for a single universe-tree, because the present is defined as the branching point which is indeed uniquely defined on a single universe-tree [p. 346]:

> Given that the universe-picture for 1976 has the form of a tree, it is not arbitrary how we select the tree for the year 2000. In fact, we cannot select it at all. If we recall that there is no distinguished future branch, we see that there is no way in which the universe-picture for 2000 can be constructed in 1976.

But how do we know which universe-tree to select from the whole array of universe-trees in order to determine the 'real' present. McCall has to provide an objective method for determining the 'latest' universe-tree in the whole array of universe-trees, which means that he has to give objective criteria for determining that universe-tree which is itself not a proper subtree of any other universe-tree. It appears, however, that in order to explain within the context of McCall's theory that the years 2000 and 1800 cannot constitute a present, it is necessary to refer to a conscious observer living between 1800 and 2000 [McCall 1976, p. 347]:

> The year 2000, for example, cannot constitute a 'present', since there exists no means of determining (other than waiting and seeing) which universe-tree it serves to divide. Similarly, the year 1800 cannot serve as 'present', since not all of the branches of its universe-tree are on a par: one and only one of them is distinguished as the future.

This is true, given the arbitrary assumption made by McCall that the universe-picture of 1976 has the form of a tree. Evidently, this assumption is not sufficient to guarantee an objective determination of a unique moment as the present. But without this assumption it is not possible to determine objectively a unique moment of time as the present, for then the foregoing is only true if a tacit appeal is made to a conscious observer living some time between 1800 and 2000. Only for such an observer it is necessary to 'wait' until the year 2000 to see which universe-tree that moment will serve to divide, and only for such an observer it is possible to distinguish one of the branches of the universe-tree of 1800 as 'the' future.

In the above interpretation of McCall's theory, the world is modelled after a deck of cards, each card containing a universe-picture at a certain time. We have reached the conclusion that starting from such interpretation, McCall's theory does not justify the conclusion that the flow of time is an aspect of physical reality. However, it appears that McCall would object to the above interpretation of his theory. In his correspondence with Smart [Smart 1980] he has stated that it somehow misses an important aspect of his theory. For McCall, the universe is not like a deck of cards, but like one card: 'The universe at time t consists of just the universe at t, and the universe at t', where t' is earlier or later than at t, does not exist at all.' This means that at each moment just one card of the deck exists, and the present corresponding to the universe-picture on that card, constitutes the real, unique present. But it is difficult to understand why such a characterisation of the present would be objective. If the unique present is singled out by an ontological requirement , viz., the existence of the corresponding universe-picture (card), why would it be necessary to develop a theory like McCall's? Ultimately, this characterisation of the present is the same as the one given under point 3 at page 200. This characterisation, however, is not objective, for McCall does not make clear how his ontological underpinning of the flow of time can be stated in purely physical terms.

Let us now see whether the minddependent theories of time flow can do any better.

According to the minddependent theories, the flow of time is a 'subjective' phenomenon in the sense that it does not figure in any description of the objective physical world. The flow of time is considered to play a role only with regard to the experience of a knowing subject (a mind, a conscious observer).

It will be helpful to distinguish between a 'weak' and a 'strong' version of the minddependency of time flow. The weak version maintains that, although the flow of time is not an aspect of physical reality, it is nevertheless a real irreducible aspect of the mind. The strong version goes further in that it maintains that even within the context of the experience of a conscious observer, the flow of time is not a genuine phenomenon, but is only an illusion (5). Grünbaum's minddependent theory of time flow, to be discussed below, exemplifies the weak version, whereas Parmenides' view belongs to the strong kind.

Several reasons in favour of the minddependency of the flow of time can be given. In the first place, it is extremely difficult to give a precise meaning to the expression 'flow of time'. The expression itself is nothing but a (kinematical) metaphor based on our intuitive picture of time as a straight line. In this picture, the flow of time is represented by the motion of the now in the direction of the future. However, the motion of this now cannot be analysed or understood relatively to time itself, but requires something like a second time coordinate. One also quickly ends up asking 'how fast' the now is moving along the time axis. Thus, a too literal interpretation of this metaphor leads immediately to great difficulties.

Likewise, it can be shown that phrases such as "the present advances into the direction of the future" or "time flows into the direction of the future" are mere tautologies (similar to statements such as "all bachelors are males") if the meaning of 'advancing' or 'flowing' in a temporal direction is construed analogically to the meaning of the same terms in the case of a spatial direction [cfr. Grünbaum 1971, p. 202-203]. Such expressions therefore do not convey any information about the meaning of the notion of the flow of time.

Thus, it proves extremely difficult to get a hold on the notion of flow of time. This being the case, and realising that physics does get along very well without introducing such concepts - which is rather remarkable - it is no great wonder that physicists are inclined to degrade this flow of time into an illusion, or to consider it to be minddependent.

Finally, the rare attempts at constructing an objective theory of time flow have not been very successful until now, as we have seen just above.

In the light of the foregoing, it is not surprising that attempts have been made to develop and defend the idea that time flow is minddependent [Grünbaum 1971,1973; Christensen 1976]. I shall restrict myself to a discussion of Grünbaum's defense of the minddependency of time flow, which clearly illustrates this point of view and the kind of problems it raises.

According to Grünbaum, any attempt to base an objective theory of time flow on the fact that we live in an indeterministic universe is bound to fail because, and here I fully agree with him, there is no intrinsic connection between the flow of time and the (in)determinism of the physical world [Grünbaum 1971,p. 219 ff, 1973, p.321]. It is not possible to define the now as the moment at which that which was not determined becomes determined. Therefore Grünbaum asks himself in what other way the occurrence of a physical event can rightly be qualified as occurring now. His answer is that an appeal to a 'mind-possessing organism' is inevitable in qualifying an event as occurring now. Becoming is therefore minddependent [Grünbaum 1971, p. 197]:

> Becoming is minddependent because it is not an attribute of physical events per se but requires the occurrence of certain conceptualized conscious experiences of the occurrence of physical events.

More in particular, the 'nowness' of an event is not an attribute of physical events [Grünbaum 1971, p.206]:

> what qualifies a physical event at a time t as belonging to the present or as now is not some physical attribute of the event or some relation it sustains to other purely physical events. Instead what is necessary so to qualify the event is that at the time t at least one human or other mind-possessing organism M is conceptually

> aware of experiencing at that time either the event itself or another event simultaneous with it in M's reference frame.

Thus, not only awareness, but conceptualized awareness or judgmental awareness is required. It is not sufficient that a person experiences a certain event at time t, but it is also necessary that he knows that he is experiencing that event.

In commenting on his characterisation of the now, Grünbaum remarks that the claim that the now is minddependent does not assert that the nowness of an event is arbitrary [Grünbaum 1971, p. 210]: "it is not at all arbitrary what event or events qualify as being now at any given time t". In this respect, however, there is an important difference between Grünbaum's minddependent theory of time flow and the objective theories discussed above. The latter seek to determine on objective grounds a unique moment of time as the now. Grünbaum on the other hand only describes what qualifies events as being now at a time t whatever moment t is. So, no unique moment of time is singled out as the present.

Here we touch upon the first serious shortcoming of Grünbaum's theory, for any theory about the flow of time, which does full justice to the role of time in our experience, has to account for the fact that just one privileged moment of time is called 'the present'.

But there is a second difficulty. A peculiar feature of Grünbaum's theory is that all events fall apart into two classes: mental events that do become and physical events without becoming. This poses an intriguing problem [Grünbaum 1971, p.216]:

> Whence the becoming in the case of mental events, given that physical events themselves do not become independently of being perceived but occur tenselessly? More specifically, the question is: if our experiences of (extra and/or intradermal) physical events are causally dependent upon these events, how is it that the former mental events can properly qualify as being 'now', whereas the eliciting physical events themselves do not so qualify, and yet both kinds of events are (severally and collectively) alike

> related by quasi-serial relations of earlier and later?

In other words, how can the becoming of mental events which are causally dependent upon physical events, be reconciled with the non-becoming of the physical events? According to Grünbaum, this problem dissolves as soon as we recognise that complex mental states have distinctive features of their own (in casu: the attribute of nowness) and that any account of the 'place of mind in nature' has to take due notice of 'what may be peculiar to the time of awareness' [ibidem, p.216]. So his reply to the above question is:

> Becoming can characterize mental events qua their being both bits of awareness and sustaining relations of temporal order.

Grünbaum's reply, however, is not very convincing. On the one hand, he assumes that the flow of time is not a property of the physical world whereas, on the other hand, he assumes that the physical events are nevertheless temporally ordered (in some 'tenseless' sense).

I shall now turn to some of the main problems inherent to the objective and minddependent theories of time flow. My intention is to show that we are confronted with a serious dilemma.

First, the objective theories of time flow. In my opinion, it is doubtful whether the attempts to construct an objective theory of time flow have any chance of success. In studying physical reality, physicists concentrate upon reproducible phenomena, and they eliminate all that makes a phenomenon unique. In particular, they abstract from the fact that an event takes place 'here and now'. But whereas physics generally tries to describe the universal aspects of the phenomena, the goal of an objective theory of time flow is precisely to get hold of the unique: such theory must single out a unique moment of time as the present which separates the past from the future. Therefore, it is in principle questionable whether an objective theory of time flow is feasible. It appears that the universality of the laws of physics precludes the introduction of the notion of the present, and consequently also of the notion of the flow of time, on the basis of these laws. Proponents of objective theories of time flow must find a way to deal with this problem. I see no way of disarming this argument.

But in my opinion, the minddependent theories are not better off.

In the first place, the minddependent theories of time flow lead to the awkward conclusion that, from a physical point of view, there is (almost) no difference between space and time. Consider the Newtonian picture of time; if the flow of time is not part of the structure of physical time, then what is left of Newton's conception of time is just a straight line, since the motion of the present along this line has disappeared. In that case, space and time would differ structurally only with respect to their number of dimensions, space being three-dimensional and time one-dimensional. But clearly, this is not the most significant difference between space and time. What really distinguishes the two are the notions of past, present and future (and thus the notion of the flow of time), which have no counterpart for space. But precisely, these would be physically irrelevant according to the minddependent theories of time flow, and thus time would become almost completely "spatialised" [cfr. Smart 1955].

This problem becomes even more acute within the spacetime description of physical reality. Through the unification of space and time into a four-dimensional spacetime manifold by means of the universal constant of homogenisation c (= the velocity of light in vacuo), the specific character of time is completely lost. Apart from the different sign in the spacetime metric, the time coordinate is structurally indistinguishable from the space coordinates. As remarked above, only the introduction of the notions of past, present, future and flow of time would turn coordinate 'time' into a real time. If these notions too have no objective significance with regard to parameter time, then it is not clear in what sense space and time are different.

Therefore I cannot agree with Grünbaum when he states that, in spite of the minddependency of the flow of time, the physical events themselves are temporally ordered. It is not sufficient to qualify the formal ordering of the events on a worldline as temporal, just because of the fact that this ordering is invariant whereas no such invariant ordering exists for the spatial betweenness [Grünbaum 1971,p. 198]. In my opinion, it is not possible to assert, as does Grünbaum, that on the one hand, physical events do not become independently of being perceived, and that on the other hand physical events themselves are nevertheless 'temporally' ordered without being perceived. According to Grünbaum, physical events occur tenselessly; then he assumes, without any justification, that this tenseless occurrence leads to a temporal ordering of the physical events themselves. But it is not clear at all that the occurrence of physical events in the tenseless sense can generate a real temporal ordering. This tenseless occurrence of events only leads to a formal ordering relation between the physical events, not to a temporal ordering.

If indeed the minddependency of time flow leads to the conclusion that physically there is no difference between space and time (in the sense that there is no reason to give a different semantics to the formal ordering relations between the events in a spacetime manifold: in the one case spatial, in the other temporal) then the proponents of the minddependency of time flow have to face the consequences of this conclusion for the physicist's conception of nature. It appears that, e.g., it would be necessary to review the notion of causality which is traditionally closely connected with the notion of time, for instance in relativity theory.

The second issue with regard to the minddependent theories which deserves our attention concerns the fact that the flow of time goes into the 'same direction' for each person (mind). If we want to avoid the use of the rather suspicious kinematical metaphor, then the problem involved is the following: how can we explain that the minddependent transiency of the now allows for a coherent classification of the events into past, present and future for different observers, keeping in mind that this classification is continually changing because of the transiency of the now. In other words, why is it not possible for a certain event to be in my past and in the future of somebody else, and why are opposite time flows excluded?

According to Grünbaum, this parallellism in the flow of time is related to the parallel increase in entropy in branch systems. In short, his reasoning runs like this. We experience the forward flow of time in the direction in which the information stored in our memory increases [Grünbaum 1973, p. 289]. But the physical processes underlying the formation of memories in our brains can be analysed in terms of branch systems and entropy increase [ibidem]:

> We saw that the production of traces, records, or memories is usually accompanied by entropy increases in the branch systems. And therefore it follows that the direction of increase of stored information or 'memories' either in inanimate recording-devices or in memory-gathering organisms like man must be the same as in the direction of entropy increase in the majority of branch systems.

And thus, Grünbaum concludes that "what is psychologically later goes hand in hand whith what is purely physically later on the basis of the entropic evolution of branch systems" [op. cit. p 289]. It is rather remarkable that Reichenbach, a defender of the objective theory of time flow, uses a similar argument in order to explain the parallellism in our subjective experience of the flow of time [Cfr. Reichenbach 1971, p. 269-270].

For Grünbaum, therefore, the parallellism in the transiency of the minddependent now is grounded in physical reality. But it is hard to understand how the possibility of a coherent classification of the events into past, present and future (and the usefulness of expressions such as 'the becoming of events' and 'the flow of time') finds its origin in physical reality for which there is no becoming and no flow of time. Somehow, Grünbaum thinks that the gap between the physical and the mental can be bridged by relating certain aspects of the mental to the physical world. Nevertheless, mental and physical events belong to different categories, for only mental events are characterised by a flow of time and becoming. Here we see that the problem of the minddependency of becoming ties in with the more general mind-body problem.

At this point, it is appropriate to end this inquiry into the structure and role of physical time. It appears that the language of physics is not able to deal adequately with the problem of the flow of time, at least at the present moment.

NOTES

1
This epilogue is a slightly modified version of Kroes [1984].
2
For a discussion of Aristotle's ideas on time, see Owen [1976].
3
The distinction between mechanical, evolutive, historical and individual time is partly based upon Scheurer [1979, Ch. 12] and Scheurer [1984]. There are also strong similarities with Denbigh [1981].
4
With regard to traces, it appears to be possible to give a physical foundation to the fact that we can have traces of the past but not of the future. This asymmetry in traces, however, is not really an aspect of the flow of time, but is related to time asymmetry for it is based upon a fundamental asymmetry between prediction and retrodiction [Kroes 1983a].
5
The explanation of the origin of this illusion poses a special problem; cfr. Hinckfuss [1975] and Smart [1980]. Although Smart considers the flow of time to be an illusion, he does not adhere to the strong version of the minddependency of time flow. In correspondence with the author he pointed out that this version presupposes that there exists a non-physical mind, whereas he rejects this assumption (cfr. also Smart [1949]).

BIBLIOGRAPHY

(Occasionally the first date of appearance of a book is given in brackets at the end of a reference.)

-Abramenko, B. [1958]: "On dimensionality and continuity of physical time and space", Britisch Journal for the Philosophy of Science, vol IX, no 34, pp 89-109.

-Alexander, H.G. [1956] (ed.): The Leibniz-Clarke correspondence, Manchester Univ. Press, Manchester.

-Aristotle: The works of Aristotle, vol II 'Physica', (W.D. Ross, ed.), At the Clarendon Press, Oxford (1930).

-Bergson, H. [1944]: Creative evolution, Random House, New York.

-Bohm, D. [1965]: The special theory of relativity, Benjamin, New York.

-Boltzmann, L. [1964]: Lectures on gas theory, transl. S.G. Brush, Univ. Calif. Press, Berkeley (1896, 1898).

-Bondi, H. [1967]: Assumption and myth in physical theory, Cambridge Univ. Press, Cambridge.

-Brush, S.G. [1976]: The kind of motion we call heat: a history of the kinetic theory of gases in the 19th century, 2 vols, North-Holland Publ Co.

-Bunge, M. [1968]: "Physical time: the objective and relational theory", Philosophy of Science, 35, pp 355-388.

-Cartan, E. [1923]: Ann. Ecole Norm., 40, 325.

-Cartan, E. [1924]: Ann. Ecole Norm., 41, 1.

-Choquet-Bruhat, Y. et al. [1977]: Analysis, manifolds and physics, North Holland Publ. Co., Amsterdam.

-Christensen, F. [1976]: "The source of the river of time", Ratio, 18, pp 131-144.

-Christensen, F. [1981]: "Special relativity and spacelike time", Britisch Journal for the Philosophy of Science, 32, pp 37-53.

-Christenson, J.H. et al. [1964]: "Evidence for the 2π decay of the K^0 neutral meson", Physical Review Letters, vol 13, 138.

-Copleston, F. [1962]: A history of philosophy, vol 1 'Greece and Rome', part I, Image Books, New York.

-Davies, P.C.W. [1977]: The physics of time asymmetry, Univ. Calif. Press, Berkeley.

-Denbigh, K. [1981]: Three concepts of time, Springer Verlag, Berlin.

-Dirac, P.A.M. [1926]: "Quantum mechanics and a preliminary investigation of the hydrogen atom", Proceedings Royal Society London (A), 110.

-Dorling, J. [1970]: "The dimensionality of time", American Journal of Physics, vol 38I, pp 539-540.

-Earman, J. [1970]: "Space-time, or how to solve philosophical problems and dissolve philosophical muddles without really trying", Journal of Philosophy, vol LXVII, no 9, pp 259-277.

-Earman, J. [1970a]: "Who's afraid of absolute space?", Australasian Journal of Philosophy, vol 48, no 3, pp 287-319.

-Earman, J. [1972a]: "Some aspects of general relativity and geometrodynamics", Journal of Philosophy, vol 69, pp 634-647.

-Earman, J. [1972b]: "Notes on the causal theory of time", Synthese, 24, pp 74-86.

-Earman, J. [1974]: "An attempt to add a little direction to the "problem of the direction of time"", Philosophy of Science, 41, pp 15-47.

-Earman, J. et al. [1977a] (eds.): Foundations of spacetime theories, Minnesota Studies in the Philosophy of Science, vol VIII, Univ. of Minnesota Press, Minneapolis.

-Earman, J. [1977b]: "How to talk about the topology of time", Nous, 11, pp 211-226.

-Eddington, S.A. [1928]: The nature of the physical world, The McMillan Co, New York.

-Ehrenfest, P. and T. [1959]: The conceptual foundations of the statistical approach in mechanics, Cornell Univ. Press, Ithaca.

-Einstein, A. [1905]: "Zur Elektrodynamik bewegter Körper", Annalen der Physik, 17. Translated in Einstein [1952].

-Einstein, A. et al. [1952]: The principle of relativity, A collection of original memoirs by Lorentz, Einstein, Minkowski and Weyl; Dover Publ., New York (1923).

-Einstein, A. [1954]: Relativity, the special and the general theory, Methuen and Co., London (1916).

-Feynman, R. [1973]: The character of physical law, M.I.T. Press, Cambridge, Mass. (1965).

-Flint, H.T. [1948]: "The quantization of space and time", Physical Review, vol 74, no 2, pp 209-210.

-Frank, P. [1909]: Sitzber. Kgl. Akad. Wiss. Wien, Math. Naturw. Kl 118, abt. IIa, 373.

-Fraser, J.T. et al. [1978] (eds.): The study of time III, Springer Verlag, Berlin.

-Friedrichs, K. [1927]: Math. Ann., 98, 566.

-Geroch, R. [1979], G.T. Horowitz: "Global structure of spacetimes", in Hawking [1979].

-Gold, T. [1962]: "The arrow of time", American Journal of Physics, vol 30, pp 403-410.

-Gold, T. [1967] (ed.): The nature of time, Cornell Univ. Press, Ithaca.

-Graves, J.C. [1971]: The conceptual foundations of contemporary relativity theory, M.I.T. Press, Cambridge, Mass.

-Grünbaum, A. [1971]: "The meaning of time", in Basic issues in the philosophy of time, E. Freeman, W. Sellars (eds.), Open Court, Lasalle.

-Grünbaum, A. [1973]: Philosophical problems of space and time, Reidel Publ. Co, Dordrecht; second enlarged edition (1963).

-Hanson, N. [1958]: Patterns of discovery, Cambridge Univ. Press, Cambridge.

-Havas, P. [1964]: "Four-dimensional reformulations of Newtonian mechanics and their relation to the special and general theory of relativity", Reviews of Modern Physics, vol 36, no 4, pp 938-965.

-Hawking, S.W. [1968]: "The existence of cosmic time functions" Proceedings of the Royal Society A, 308, pp 433-435.

-Hawking, S.W. [1973], G.F.R. Ellis: The large scale structure of space-time, Cambridge Univ. Press, Cambridge.

-Hawking, S.W. [1979], W. Israel (eds.): General Relativity, Cambridge Univ. Press, Cambridge.

-Heisenberg, W. [1927]: "Ueber den anschaulichen Inhalt der quantentheoretischen Kinematik und Mechanik", Zeitschrift für Physik, 43, 172.

-Hinckfuss, I. [1975]: The existence of space and time, Clarendon Press, Oxford.

-Hoffman, B. [1966] (ed.): Perspectives in geometry and relativity, Indiana Univ. Press, Bloomington.

-Hurewicz, W. [1941], H. Wallman: Dimension Theory, Princeton Univ. Press, Princeton.

-Iyanaga et al. [1977]: Encyclopedic dictionnary of mathematics, 2 vols, Cambridge Univ. Press, Cambridge, Mass.

-Jammer, M. [1974]: The philosophy of quantum mechanics, Wiley and Sons, New York.

-Kant, I. [1956]: Kritik der Reinen Vernunft, Felix Meiner Verlag, Hamburg (1781).

-Kant, I. [1969]: Prolegomena zu einer jeden künftigen metaphysik, Felix Meiner Verlag, Hamburg (1783).

-Klein, F. [1921]: "Erlanger Programm", Ges. Math. Abh. Ie band, Springer Verlag, Berlin.

-Kline, M. [1972]: Mathematical thought from ancient to modern times, Oxford Univ. Press, New York.

-Koyré, A. [1965]: Newtonian Studies, Univ. Chicago Press, Chicago.

-Kroes, P.A. [1982]: "Order and irreversibility", Nature and System, 4, pp 115-129.

-Kroes, P.A. [1982a]: An inquiry into the structure of physical time, thesis, Catholic Univ. Nijmegen, The Netherlands.

-Kroes, P.A. [1983]: "The clock paradox, or how to get rid of absolute time?", Philosophy of Science, 50, no 1, pp 159-163.

-Kroes, P.A. [1983a]: "Traces, Prediction and Retrodiction", Nature and System, vol 5, pp 131-146.

-Kroes, P.A. [1984]: "Objective versus minddependent theories of time flow", Synthese, in the press.

-Kronheimer, E.H. [1971]: "Time-ordering and topology", General relativity and gravitation, vol 1, no 3, pp. 261-268.

-Kuhn, T.S. [1978]: Black-body theory and the Quantum Discontinuity, 1894-1912, Clarendon Press, Oxford.

-Kuznetsov, I.V. et al. [1965] (eds.): Philosophical problems of elementary particle physics, Israel progam for scientific translations, Jerusalem.

-Lakatos, I. [1978]: Philosophical Papers, 2 vols, (J. Worrall and G. Currie eds.), Cambridge Univ. Press, Cambridge.

-Lanczos, C. [1962]: "The splitting of the Riemann tensor", Reviews of Modern Physics, vol 34, no 3, pp 379-389.

-Landau, L.D. [1974], E.M. Lifschitz: Lehrbuch der theoretischen Physik, vol III 'Quantenmechanik', Akademie Verlag, Berlin.

-Landsberg, P.T. [1970]: "Time in statistical physics and special relativity", Studium Generale, 23, pp 1108-1158.

-Landsberg, P.T. [1982]: The enigma of time, Adam Hilger Ltd, Bristol.

-Laudan, L. [1977]: Progress and its problems, Univ. Calif. Press, Berkeley.

-Lautman, A. [1977]: "Le problème du temps", in Essai sur l'unitè des mathèmatiques et divers ècrits, Union Gènèral d'èditions 10/18 (1946).

-Layzer, D. [1971]: "Cosmogonic Processes", in Astrophysics and General Relativity, vol 2, (Chretien, Deser and Goldstein, eds.), Gordon and Breach, New York, pp 151-233.

-Layzer, D. [1972]: "Cosmology and the arrow of time", in Vistas in astronomy, vol 13, (A. Beer, ed.), Pergamom Press, Oxford, pp 279-287.

-Layzer, D. [1975]: "The arrow of time", Scientific American, vol 223, no 6, pp 56-69.

-Lucas, J.R. [1973]: A treatise on time and space, Methuen, London.

-Mandelbrot, B.B. [1977]: Fractals, Freeman Co, San Francisco.

-Matthews, G. [1979]: "Time's arrow and the strucure of spacetime", Philosophy of Science, 46, pp 82-97.

-McCall, S. [1976]: "Objective time flow", Philosophy of Science, 43, pp 337-362.

-McCarty, G. [1967]: Topology, McGraw-Hill Book Co, New York.

-Mehlberg, H. [1980]: Time, causality, and the quantum theory, vol I, (R.S. Cohen, ed.), Reidel Publ. Co, Dordrecht.

-Mehra, J. [1973] (ed.): The physicist's conception of nature, Reidel Publ. Co, Dordrecht.

-Milne, E.A. [1948]: Kinematic Relativity, Oxford Univ. Press, Oxford.

-Minkowski, H. [1908]: "Space and time", in Einstein [1952].

-Misner, C.W. et al. [1973]: Gravitation, Freeman and Co, San Francisco.
-Mittelstaedt, P. [1976]: Der Zeitbegriff in der Physik, Bibliographisches Institut, Mannheim.
-Neumann, J. von [1955]: Mathematical Foundations of Quantum Mechanics, Princeton Univ. Press, Princeton.
-Newton, I. [1974]: Mathematical principles of natural philosophy, Univ. Calif. Press, Berkely (1686).
-Newton-Smith, W.H. [1980]: The structure of time, Routledge and Kegan Paul, London.
-O'Raifeartaigh, L. [1972] (ed.): General Relativity, Clarendon Press, Oxford.
-Overseth, O.E. [1969]: "Experiments in time reversal", Scientific American, vol 221, no 4, pp 89-101.
-Owen, G. [1976]: "Aristotle on time", in Motion and time, space and matter, (P. Machamer, R. Turnbull eds.), Ohio State Univ. Press, pp 3-27.
-Penney, R. [1965]: "On the dimensionality of the real world", Journal of Mathematical Physics, vol 6, no 11, pp 1607-1611.
-Penrose, R. [1979]: "Singularities and time-asymmetry", in Hawking [1979].
-Poincaré, H. [1952]: Science and Hypothesis, Dover Publ., New York (1902).
-Poincaré, H. [1958]: The value of Science, Dover Publ., New York (1905).
-Popper, K. [1972]: Objective Knowledge, At the Clarendon Press, Oxford.
-Prigogine, I. [1973]: "Time, irreversibility and structure", in Mehra [1973], pp 561-593.
-Prigogine, I. [1980]: From being to becoming, Freeman and Co, San Francisco.
-Prokhovnik, S.J. [1967]: The logic of special relativity, Cambridge Univ. Press, Cambridge.
-Rankin, B. [1965]: "Quantum mechanical time", Journal of Mathematical Physics, vol 6, no 7, pp 1057-1071.
-Reichenbach, H. [1957]: The philosophy of space and time, Dover Publ., New York (1927).
-Reichenbach, H. [1971]: The direction of time, Univ. Calif. Press, Berkeley(1956).
-Rietdijk, C. [1966]: "A rigorous proof of determinism derived from the special theory of relativity", Philosophy of Science, 33.

-Sakurai, J.J. [1973]: Advanced quantum mechanics, Addison-Wesley Publ. Co, Reading, Mass.

-Scheurer, P.B. [1970], E.C.G. Stueckelberg: "Charge conjugation C and proper time reversal", Helv. Phys. Act., 43, 738.

-Scheurer, P.B. [1973]: "Présentation d'un nouveau paradigme pour la dynamique relativiste quantique", CR des Séances, SPHN Genève, NS, vol 8, fasc. 1-3, p. 32-37.

-Scheurer, P.B. [1977]: "Vues nouvelles sur la science du temps et l'histoire", in Mélanges Jeanne Hersch, L'âge d'homme, Lausanne, pp 157-172.

-Scheurer, P.B. [1979]: Révolutions de la science et permanence du réel, Presse Universitaire de France, Paris.

-Scheurer, P.B. [1982]: "Relativization: a structural reinterpretation of the special relativity", Arch. Sc. Genève, vol 35, fasc 2, pp 217-226.

-Scheurer, P.B. [1984]: "Mouvement, Evolution, Histoire: sur trois modalités du temps physique", Revue Philosophique de Louvain, Tome 82.

-Schilpp, P.A. [1969] (ed.): Albert-Einstein: Philosopher-Scientist, Library of living Philosophers, vol VII, Open Court, La Salle (1949).

-Schlegel, R. [1977]: "The clock paradox: some new thoughts", Philosophy of Science, 44, pp 306-312.

-Scott, W.P. [1967]: Erwin Schrödinger; an introduction to his writings, Univ. Mass. Press, Mass.

-Shoemaker, S. [1969]: "Time without change", Journal of Philosophy, vol LXVI, no 12, pp 363-381.

-Sklar, L. [1977]: Space, Time and Spacetime, Univ. Calif. Press, Berkeley.

-Smart, J. [1949]: "The river of time", Mind, 58, pp 483-494.

-Smart, J. [1955]: "Spatialising time", Mind, 64, pp 239-241.

-Smart, J. [1980]: "Time and Becoming", in Time and Cause, (P. v. Inwagen ed.), Reidel Publ. Co, Dordrecht.

-Smoluchowski, M. von [1915], Wiener Berichte, 124, 339.

-Stegmüller, W. [1975]: Hauptströmungen der Gegenwartsphilosophie, Alfred Kröner Verlag, Stuttgart.

-Stein, H. [1967]: "Newtonian space-time", Texas Quarterly, X, pp 174-200.

-Stueckelberg, E.C.G. [1974], P.B. Scheurer: Thermocinètique phènomènologique galilèenne, Birkhäuser Verlag, Basel.

-Thirring, W. [1977]: Lehrbuch der Mathematischen Physik, vol 1, Springer Verlag, Berlin.

-Webb, C.W. [1977]: "Could space be time-like?" Journal of Philosophy, vol LXXIV, no 8, pp 462-474.

-Weinberg, S. [1978]: The first three minutes, Fontana/Collins, Glasgow.

-Weingard, R. [1979]: "Review of P.C.W. Davies, Space and time in the modern universe", Nature and System, vol 1, pp 60-70.

-Weingard, R. [1981]: "Review of R. Schlegel, Superposition and Interaction", Nature and System, 3, pp 243-251.

-Weyl, H. [1952]: Space, Time, Matter, Dover Publ., New York (1921).

-Weyl, H. [1963]: Philosophy of mathematics and natural science, Atheneum, New York (1927).

-Whitrow, G.J. [1972]: The nature of time, Penguin Books, Harmondsworth.

-Winnie, J.A. [1977]: "The causal theory of space-time", in Earman [1977a].

-Woodhouse, N.M.J. [1973]: "The differentiable and causal structures of space-time", Journal of Mathematical Physics, vol 14, no 4, pp 495-501.

NAME INDEX

SUBJECT INDEX

GPSR Compliance

The European Union's (EU) General Product Safety Regulation (GPSR) is a set of rules that requires consumer products to be safe and our obligations to ensure this.

If you have any concerns about our products, you can contact us on ProductSafety@springernature.com

In case Publisher is established outside the EU, the EU authorized representative is:

Springer Nature Customer Service Center GmbH
Europaplatz 3
69115 Heidelberg, Germany

Batch number: 10607222

Printed by Printforce, the Netherlands